Urbanization and Development

Brown University Studies in Population and Development

Urban Migrants in Developing Nations: Patterns and Problems of Adjustment, edited by Calvin Goldscheider

Rural Migration in Developing Nations: Comparative Studies of Korea, Sri Lanka, and Mali, edited by Calvin Goldscheider

Why People Intend to Move: Individual and Community-Level Factors of Out-Migration in the Philippines, Sun-Hee Lee

Social Change and Family Processes: Arab Communities in Shefar-A'm, Majid Al-Haj

Rural Development and Migration: A Study of Family Choices in the Philippines, Sally E. Findley

Scandinavian Exodus: Demography and Social Development of 19th-Century Rural Communities, Briant Lindsay Lowell

Urbanization and Development: The Rural-Urban Transition in Taiwan, Alden Speare, Jr., Paul K.C. Liu, and Ching-lung Tsay

Brown University Studies in Population and Development

Published in cooperation with the Population Studies and Training Center Brown University

---

Urbanization and Development

The Rural-Urban Transition in Taiwan

Alden Speare, Jr., Paul K.C. Liu, and Ching-lung Tsay

Routledge
Taylor & Francis Group
LONDON AND NEW YORK

First published 1988 by Westview Press

Published 2018 by Routledge
52 Vanderbilt Avenue, New York, NY 10017
2 Park Square, Milton Park, Abingdon, Oxon OX14 4RN

Routledge is an imprint of the Taylor & Francis Group, an informa business

Library of Congress Cataloging-in-Publication Data
Speare, Alden.
Urbanization and development.
(Brown University studies in population and development)
Bibliography: p.
1. Urbanization—Taiwan. 2. Rural development—Taiwan. I. Liu, Paul K.C. II. Tsay, Ching-lung. III. Title. IV. Series.
HT147.T3S64 1988 307.7'6'0951249 87-29532

ISBN 13: 978-0-367-21275-9 (hbk)
ISBN 13: 978-0-367-21556-9 (pbk)

Contents

Tables

Figures

Foreword

The growth and expansion of cities and the transition from a rural to an urban society are among the most critical links between population change and economic development. On the one hand, migration is one of the fundamental demographic processes associated with changes in the population of urban places; the changing distribution of population between urban and rural places has been associated in the West historically with the revolutions in mortality and fertility control. On the other hand, industrialization and economic expansion emerged in the cities of Western nations, bringing new opportunities and challenges to the welfare of nation states and their populations.

This volume examines these patterns of demographic and economic change in developing nations by focusing on urbanization and migration in Taiwan, and by investigating the linkages between these processes and increases in population size and growth of the economy. Over the last several decades, Taiwan has been transformed from a rural to an urban society and has undergone the demographic transition from high to low birth and death rates. Estimates for the mid-1980s suggest that Taiwan is characterized by an expectation of life of 73 years, rates of infant mortality less than 10 per 1,000 live births, and a total fertility rate of only 1.9 children per woman. Fully two-thirds of its population live in places designated as urban. These demographic indicators place Taiwan among the most developed nations of the world.

Along with these demographic and urban transitions, the economy has shifted over the last several decades from agriculture to industry, from local to international trade, and from relatively simple to more complex economic specialization and diversification. These changes gained momentum after World War II and have dramatically altered the lives of the population living in Taiwan. Standards of living have improved, educational levels have increased, intergenerational occupational mobility has intensified, and processes of modernization have clearly emerged. Yet, Taiwan has retained some unique features including both its social organization and culture, and its international linkages, that distinguish it from other societies.

What determined the rates of urbanization in Taiwan? What are the consequences of population concentration in the cities and metropolitan areas of Taiwan? What are the ways in which economic development and population transformation are linked? How are these broader societal trends translated into changing lifestyles and living standards of the people living in Taiwan? These are the questions that Alden Speare Jr., Paul K.C. Liu and Ching-lung Tsay, address in this monograph. They approach their subject with an authoritative, in-depth, sophisticated investigation of the major contours of social, economic, and demographic change. Their detailed knowledge of Taiwan, its history and society, does not prevent them from seeing the broader implications of their case study. They compare and contrast demographic and economic patterns in Taiwan to those found in research on urbanization, migration and development in other countries. In turn, they generate important new research hypotheses and point to policy implications for other developing nations of the Third World. Their comparisons enhance the quality and value of their case study.

Nevertheless, their attention is primarily directed to Taiwan. Using a wide range of data sources--censuses, population registration data, and sample survey data sets--the authors expertly guide us through an assessment of social, economic, and demographic changes in Taiwan over the last decades in the context of the rural-urban transformation and the development of new city configurations. Details on urbanization in Taiwan as a whole and on the expansion of Taipei are analyzed in combined economic and demographic contexts.

The demographic transformation in Taiwan is usually attributed to the development of an extensive family planning program and the debate has centered over its role in the decline in fertility. In this volume, Speare and his colleagues have documented the central role of another demographic process in the socio-economic development of Taiwan, migration. Movement from rural areas to urban places was part of the economic development and the demographic revolutions that characterized Taiwan. The importance of migration in the development-demographic relationship finds extensive support in this analysis of Taiwan and is applicable more widely to other developing nations.

The book focuses on both macro- and micro-level issues. Starting at the macro level, the volume provides an historical overview of urban growth in Taiwan, pointing out the relatively smooth distribution of cities of various sizes and the absence of disproportionate population concentration, in a primate city. Using the wealth of data available on urban growth, with careful attention to problems of classification and definitions over time, the authors document the emerging system of cities in Taiwan, the fit of the rank-size rule in Taiwan, and the critical role of intermediate cities in its urban development. The analysis proceeds to relate these urbanization patterns to the processes of socio-economic development in Taiwan, revealing the positive linkages between these macro processes. Cross-sectional and longitudinal analyses show the powerful connections between indicators of economic development and patterns of urbanization.

A detailed examination of the Taipei Metropolitan Area reveals how urban expansion and suburbanization characterize the largest city of Taiwan. Migration from rural areas and the natural increase of the migrants played a critical role in the urban growth and expansion in Taipei. Examining economic indicators, the authors conclude that issues of labor "underutilization" and "inadequate" employment are critical in understanding the consequences of demographic and economic changes, not simply the direction of labor force expansion or unemployment.

Migration and other demographic changes are not the only responses to economic changes and the expansion of the opportunity structure. The analysis connects migration to commuting and rural industrialization as alternative responses in the transition from an agrarian to an industrial economy.

An in-depth examination of migration requires a consideration of the determinants of rural-urban migration and an investigation of the integration of migrants in urban settings. Using extensive survey data to capture the decision-making process, the authors specify four factors that are involved in determining migration: the social and economic bonds at places of origin; residential and job satisfaction at places of destination; awareness of opportunities elsewhere; and the expected costs and benefits of the move. The specification of these determinants of migration is the basis for weaving together sociological and economic models and investigating empirically individual- and household-level decisions to move. Expanding beyond the over simplified human capital approach to migration, the authors study these factors in combination to help account not only for patterns of migration but to deal with the no less critical question of why many had not considered moving. The use of sophisticated migration models results in clear and sharp analyses. Thus, for example, they conclude that "bonds" to places of origin are important in the decision to move. They further specify that bonds to community of origin are diverse; some bonds are more likely to link people to places of origin and thereby limit migration while other bonds, particularly those that place people in subordinate positions, may facilitate migration. Clearly, these findings have implications for understanding the determinants and consequences of migration processes in other contexts and places.

Another important set of findings addresses the relative economic success of migrants, compared to those who do not move and to those who are native urbanites. Multiple comparisons of migrants over time and in relationship to those in places of origin suggest that movers are better off due to migration. Movers to urban places are particularly successful economically after five years duration of residence in the city. While new arrivals have initial difficulty finding a job, their employment prospects improve with time. In terms of economic adjustment, as well as other indicators of living conditions and social life, migration seems to have had positive consequences for the migrants and migrant families. In combination with the positive linkages between migration, urban growth, and economic development at the macro level, the emerging picture in Taiwan points unmistakably to the advantages of migration at both the individual and societal levels.

Urbanization and Development: The Rural-Urban Transition in Taiwan is the seventh volume in the Brown University series on Population and Development. The research effort that forms the backbone of this analysis has already had a major impact on theoretical, methodological, and substantive issues in the study of urbanization and the adjustment of migrants in the cities of developing nations. Several of the research projects reported in previous volumes in this series have been directly influenced by the orientation to these issues reflected in this monograph. Research continues to build on the solid foundations that Speare and his colleagues and students at Brown University and in Taiwan have established. All those who study the links between migration, urbanization, and development and who examine the adjustment of migrants in urban areas will learn from this carefully crafted case study of Taiwan. No study of the demographic transition in developing nations can ignore the critical lessons emerging from this study.

Calvin Goldscheider
Brown University

Preface

This book is the culmination of over 20 years of collaborative work by the authors. I first met Paul Liu in 1965 when we were both students in the graduate program in Population Studies at the University of Michigan. I was fortunate to be able to accompany Ronald Freedman on a two week trip to Taiwan in December 1965 and to stay for an additional week during which I travelled around the island with Paul Liu visiting villages, factories, and government offices. Paul Liu was helpful in many ways when I returned to Taiwan in 1967-68 to conduct field work for my dissertation on "The Determinants of Rural to Urban Migration in Taiwan." After completing the dissertation and receiving the Ph. D. in 1969, I continued to do research on population processes in Taiwan and I returned there during the summers of 1971, 1972, and 1973 and for several shorter visits since then.

In the summer of 1973 we conducted the migration survey which was the basis for the analysis described in chapters 6, 7 and 8. At that point, Ching-lung Tsay, who was a recent graduate of Taiwan University, was hired as one of the two field supervisors. As a tireless worker with concern for thoroughness and accuracy, Tsay added considerably to that project. In 1974, Tsay entered the graduate program at Brown University and completed a Ph.D. in 1979. Some of the main findings of his dissertation on "Employment and Earnings of City-ward Migrants: A Study of Individual Outcomes of Migration to Taipei" are summarized in chapter 7.

Parts of this monograph contain materials which have been published in Taiwan, although all of these have been revised and updated to provide a coherent study of urbanization. Chapter 2 includes a section from Paul K.C. Liu, "Labor Mobility and Utilization in Relation to Urbanization in Taiwan" *Industry of Free China* 57:5 (May 1982). Another part of chapter 2 and most of chapter 3 are revised versions of sections of a paper by Paul K.C. Liu entitled "The Relationship Between Urbanization and Socio-economic Development in Taiwan," in *Conference on Population and Economic Development in Taiwan*, Taipei: The Institute of Economics, Academia

Sinica (1976). Parts of chapter 4 were taken from Paul K.C. Liu, "Urbanization and Employment Growth in the Taipei Metropolitan Area," *Industry of Free China* (May and June 1986) and from Ching-lung Tsay, "Migration and Population Growth in Taipei Municipality," *Industry of Free China* 57:3 (March 1982). The analysis of the 1972 labor force survey data in chapter 5 is based on Paul K.C. Liu and Alden Speare, Jr., "Urbanization and Labor Mobility in Taiwan," *Economic Essays* 4 (National Taiwan University, 1973).

The 1973 Taiwan Migration Survey would not have been possible without the generous support of the Ford and Rockefeller Foundations through their program in support of Social Science Research and Legal Research on Population Policy and from the Population Council and the Joint Commission on Rural Reconstruction in Taipei. The writing of the monograph was facilitated by a grant from the Population Studies and Training Center at Brown University and a Fulbright Fellowship which enabled Ching-lung Tsay to spend time at Brown University during 1982 and 1985. Alden Speare also received support from the Faculty Development Fund at Brown University to cover some of the production costs.

The authors would like to thank Judie Quattrucci and Carol Walker for typing the manuscript and Calvin Goldscheider and Mary Speare for help in copy editing.

Alden Speare, Jr.
Brown University

1

Economic Development and Population Growth

The postwar economic development of Taiwan has been one of the real success stories in the less developed world. Starting from a level not very different from other less developed countries in the late 1940s, Taiwan had achieved a per capita GNP of $3,145 in 1985 which placed it well above the average of $1,850 for "upper-middle-income" countries according to the World Development Report (World Bank, 1987). Although there were several aspects to Taiwan's development, probably the most important was the rapid growth of manufacturing. In 1950, Taiwan was predominantly agricultural with most manufacturing centered on the processing of farm or forestry products. By 1980, one-third of the labor force was employed in manufacturing and the proportion in manufacturing exceeded that of several Western nations such as England, France, Italy, and the USSR.

The economic development and industrialization of Taiwan was accompanied by rapid urbanization. As we shall show in the following chapters, Taiwan was transformed during this period from a largely rural country to one in which the majority of the population lived in metropolitan areas or cities of over 100,000 population. While urban growth has been rapid, it has not been excessive and growth has been spread out both over time and space. While Taiwan has experienced some of the problems associated with rapid growth such as traffic congestion and air pollution in major cities, it has avoided other problems such as widespread urban unemployment and extensive squatter settlements which have plagued many less developed countries.

The relative ease with which Taiwan has progressed through the rural to urban transition, the balanced growth between different regions and between core and peripheral areas, and the adjustment of migrants to urban settings all provide examples for other less developed countries which are still mostly rural. It is our goal to describe the process of urbanization in Taiwan and to explore the determinants and consequences of urbanization both at the aggregate level and the individual level. We will seek to answer both why individual cities have grown and why individuals have moved to these places. We shall investigate the effects of growth on places and individuals. While some of the answers will relate to factors which are unique to Taiwan, many are factors which could be replicated in other settings.

Taiwan's urbanization has been shaped, to a large extent, by the course of economic development and population growth. Before examining the details of the urbanization process, it will be helpful to review the course of development and population growth. What are the factors responsible for Taiwan's rapid development? What has been the history of population growth and how has it been related to development? In the rest of this chapter, we will try to answer these questions.

THE ECONOMIC DEVELOPMENT OF TAIWAN

Taiwan's successful economic development has received widespread recognition.[1] Little (1979, p. 448) states that "Taiwan has a good claim to be ranked as the most successful of the developing countries." Since the early 1950s, Taiwan has been one of four Asian countries to experience sustained and rapid economic growth starting from a low level after World War II without reliance on substantial mineral resources.[2] However, two of the others, Hong Kong and Singapore are city states which function partly as international trade centers for products produced in other countries. The other country, Korea, has equalled Taiwan in overall economic growth, but has not done as well in distributing this growth among different regions of the country or sectors of the economy (Lau, 1986).

The growth of Taiwan's economy is briefly summarized in Table 1.1. The most impressive thing about this data series is the regularity of the growth over a long period of time. Although there have been slowdowns in growth, most notably in 1974 and 1982, when growth is averaged over five year periods and adjusted for inflation, it varies only from 6.4 to 10.6 percent per year. The most rapid growth occurred during the period from 1975 to 1980, although it also was high in the early 1950s and during the entire decade of the 1960s.

In addition to the rapid growth in GNP and GNP per capita, Taiwan accomplished a very significant transformation from an agricultural nation to an industrial nation during this period. In 1951, 32 percent of the gross national product was obtained from the primary sector of the economy

----

[1] Taiwan's development is discussed in detail in Galenson (1979), Gold (1986), Ho (1979), Lau (1986), and Kuo (1983).

[2] Little (1979) points out that Botswana, Libya, Iran, Saudi Arabia and Arab Emirates all had high rates of growth due to mineral exports.

TABLE 1.1

INDICATORS OF ECONOMIC GROWTH IN TAIWAN

Period	Average Annual Growth in GNP in Constant 1981 Dollars	Average Annual Growth of Real Per Capita GNP	Average Annual Growth in Consumer Prices	Export as Percent of GNP in Final Year
1951-55	9.7	6.0	14.3	8.3
1955-60	6.7	3.1	10.5	11.3
1960-65	9.5	6.1	2.4	18.7
1965-70	9.8	6.5	4.4	29.7
1970-75	8.8	6.7	12.2	39.5
1975-80	10.6	8.4	8.7	53.0
1980-85	6.4	4.7	3.9	54.5

Source: Calculated from data in DGBAS, Quarterly National Economic Trends, February 1986.

(farming, fishing and forestry) and only 24 percent from the secondary sector (mining, manufacturing, utilities and construction). By 1985, agriculture accounted for less than seven percent of the GNP and the secondary sector accounted for 51 percent (DGBAS, 1986, Table 6).

A third important aspect of Taiwan's development is the changes which have occurred in the income distribution. According to figures prepared by Kuo (1983) which are presented in Table 1.2, the share of income received by families in the top quintile of the income distribution declined significantly from 61 percent in 1953 to 37 percent in 1980 while that received by families in the lowest quintile increased from 3 percent to 9 percent. While the income distribution in general became much more equal, the distribution between urban and rural families followed a curvilinear pattern with increasing urbanization. Up to the early 1970s, the share ratio of income between the most rural counties and the largest cities decreased, but thereafter it increased (Kuo, 1983, pp. 109-116). This is contrary to the normal relationship whereby early development tends to increase inequality between urban and rural areas.

TABLE 1.2

DISTRIBUTION OF INCOME AMONG HOUSEHOLDS BY YEAR

Year	Lowest Quintile	Second Quintile	Third Quintile	Fourth Quintile	Highest Quintile
1953	3.0	8.3	9.1	18.2	61.4
1961	4.5	9.7	14.0	19.8	52.0
1970	8.4	13.3	17.1	22.5	38.7
1980	8.8	13.9	17.7	22.8	36.8
1984	8.5	13.7	17.6	22.8	37.4

Source: 1953 to 1980 from Kuo (1983, p. 96-97). 1984 from Taiwan Statistical Data Book 1986, pp. 59-60.

Taiwan's rapid economic growth can be divided into three phases, roughly corresponding to the periods from 1949 to 1960, 1960 to 1973, and the period from 1973 to 1985, although the transition dates for these stages are somewhat arbitrary and subject to debate (Gold, 1986). We shall refer to these as the takeoff phase (1949-60), the labor-intensive industry phase (1960-1973), and the industrial upgrading phase (1973-1985).

The Takeoff Phase

The first phase might be referred to as the takeoff phase of development. One of Rostow's criteria for takeoff is that a country without foreign aid should have net investment equal to 10 percent of the national income (Rostow, 1960)[3] Kuo (1983:6) points out that if one counts only

[3] The other criteria are "the development of one or more substantial manufacturing sectors, with a high rate of growth" and "the existence or quick emergence of a political, social and institutional framework which exploits the impulses to expansion in the modern sector" (Rostow, 1960:39 as cited in Kuo, 1983:19n). While it is much harder to fix a precise date when these conditions were met in Taiwan, they were probably met by the mid 1950s.

domestic savings as the source for investment, that this was not reached until 1963, although foreign aid and foreign investment, when added to domestic savings put Taiwan above the 10 percent level throughout the 1950s. Thus foreign aid and investment were important factors in Taiwan's early development.

The United States took a special interest in Taiwan because of its strategic position in the international relations and military strategy for East Asia in the 1950s and early 1960s. Between 1951 and 1965, the United States provided about $ 1.5 billion in economic aid to Taiwan plus additional military aid (Myers, 1986, p. 47). Between 1951 and 1961, U.S. economic aid amounted to about 37 percent of total domestic investment. After 1961, aid as a proportion of total investment declined and ceased altogether after 1968. The economic aid went particularly into supporting infrastructure such as electricity, transportation and communications. While the value of the military aid is hard to assess, if one assumes that because of fears of invasion from Mainland China that Taiwan would have built up the military with its own resources, then the U. S. military aid made it possible for more government resources to be devoted to economic development than would have been possible otherwise (Jacoby, 1966; Scitovsky, 1986, p. 143).

The second important factor in Taiwan's early development was land reform. The Land-to-the-Tiller program was initiated by the government in 1949. The program limited the amount of land which could be held by any one family to a maximum of three hectares and required the transfer of any additional land to tenants who were tilling the land, with payments spread out over ten years.[4] About 30 percent of the payment to the landowners was in terms of shares in government industries, so that the program also provided a source of nonagricultural investment (Thorbecke, 1979). As a result of this program, the proportion of cultivators who were owners increased from 36 percent in 1949 to 60 percent by 1957. While many farmers owned only small parcels, they had an added incentive to farm the land intensively and land reform is generally credited with increasing productivity per hectare and with increasing total agricultural production in Taiwan (Kuo, 1983, pp 26- 29). Taken together with other agricultural development policies, land reform helped Taiwan to increase its exports of agricultural products during the 1950s and thus earn foreign currency to pay for imports needed for development.

The third factor which was important in Taiwan's takeoff period was monetary policy. Following a period of hyperinflation in the late 1940s, when prices increased 30 fold, high priority was given to the stabilization of

----

[4] The land reform program is described in depth in Chen (1961), Koo (1968) and Yang (1970). In addition to the Land-to-theTiller program, there were also reductions in rents paid by tenants and the sale of public land to cultivators and tenants.

prices and relatively high interest rates (in real terms) were established to keep the money supply from growing too rapidly. As a result, the inflation rate from 1955 to 1960 was held to about 10 percent per year (see Table 1.1) and during the 1960s it was generally on a par with inflation rates in the United States and other Western nations.[5] This provided a good climate for investment in new businesses and the relatively high interest rates encouraged individual savings while government policies facilitated the opening of small bank accounts.

The resulting economic growth during the takeoff period was focused on import substitution, both on critical food products such as rice, on producer inputs such as fertilizer and chemicals, and on basic consumer goods such as clothing, footwear, wood products and bicycles. The government's policy of import substitution was carried out through stiff tariffs on certain imported goods and restrictions on some others. While the policy was successful in encouraging local production of these goods, Scott (1979) debates whether a policy of import substitution really helped development or whether it would have been better to promote exports at an earlier stage of development.

By the late 1950s, the government recognized the need to produce more products for export in order to expand its markets and provide a favorable trade balance. At first, exports were mostly agricultural products such as sugar, rice, bananas, teas and canned fruits and vegetables. Between 1952 and 1955, industrial products accounted for only about nine percent of exports (Wu, 1985, p. 10). However, by 1961-65, there was a noticeable increase in the proportion of industrial products to 44 percent of total exports.

The Labor-Intensive Industry Phase

While the government had already begun to promote exports in the late 1950s, this policy was strengthened in the 1960s by providing for rebates on import duties for materials used in the manufacture of exports and by several other measures.

One of the most successful methods of promoting exports was the establishment of export processing zones in Kaohsiung in 1965 and in Nantze (outside Kaohsiung) and Taichung in 1969. These zones enabled businesses to import materials free of taxes, provided a five year exemption from corporate taxes, low cost loans and a streamlining of administrative

----

[5] Between 1961 and 1971, the consumer price index in Taiwan increased by 37.8 percent compared to 35.4 percent in the United States. The wholesale price index, which is a better measure of the effect of inflation on business, increased by only 21.5 percent, compared to 16.3 percent in the United States (U.S. Bureau of the Census, 1984, DGBAS, 1986:24).

paper work. By 1972, these zones were employing about 58,000 workers, about half of whom worked in electronics (Wu, 1985). These zones were particularly attractive to the formation of joint enterprises with foreign investors who were interested in hiring unskilled workers. The majority of the workers in these zones were young females who received wages below the average for all manufacturing in Taiwan (Wu, 1985, p. 20).

In addition to the processing zones, there were several other measures to promote exports such as the establishment of bonded factories throughout Taiwan which had most of the same benefits as the factories in the zones and the general lowering of tariffs on imported goods.

By the end of this period, the contribution of exports to total output expansion had jumped from 22 percent in 1956-61 to 68 percent in 1971-76 (Kuo, 1983, p. 149). Overall growth between 1960 and 1973 averaged 9 to 10 percent per year and real wages in manufacturing more than doubled. This was a period of unprecedented economic prosperity for Taiwan. As we shall see in later chapters, it was also a period of rapid urbanization both in terms of the growth of large cities and the development of smaller cities.

The Industrial Upgrading Phase

In 1974, Taiwan suffered a recession due to the oil crisis, which resulted in rapid increases in the prices for energy sources, many other imported goods as well as shipping costs. Consumer prices rose by 47 percent and loan rates to 26 percent (Gold, 1986, p. 98). At the same time, Taiwan continued to experience rapid wage increases due to the relative shortage of labor. All of these factors increased the costs of its exports and lead to the first trade deficit since the takeoff of economic growth. It became apparent that Taiwan could not continue to remain competitive in the international market for goods manufactured by labor-intensive industries such as textiles and apparel as more and more lower wage countries began exporting these goods.

During this period, Taiwan faced international threats related to the opening up of relations between mainland China and the Western world. In the mid 1970s, while the "gang of four" was still in power in China, American businesses were threatened with loss of access to Chinese markets if they continued to do business with Taiwan and some withdrew from Taiwan operations (Gold, 1986, p. 98). While these pressures subsided after the fall of the gang of four in 1976, they made Taiwan realize the need to diversify both in terms of export products and in terms of trading partners.

The first response, which was reflected in the six-year plan for 1976-81, was to place greater emphasis on capital and technology in development. As part of this plan, there was investment in steel, petrochemicals, and other heavy industry, as well as increased investment in highways and other infrastructure. However, soon after this plan began, it was realized that there were problems with heavy industry.

The second oil crisis in 1979-80, made the high energy costs of these industries all the more apparent. There was also concern for environmental pollution, given the small land area and high population density in Taiwan. In 1977, the Economic Planning Commission which had prepared the 1976-81 plan was replaced by a new Council for Economic Planning and Development. This new Council prepared a new ten year plan for 1980-89 and a four year plan for 1982-86 which emphasized technology intensive industries instead of capital intensive heavy industry (Gold, 1986, p. 102). As one step in realizing this goal, it established a new Science-Based Industrial Park in Hsinchu which provided incentives to technology based industries while requiring them to invest in research and development (Gold, 1986, p. 103).

Steps were also taken to promote greater agricultural production. The limitation in land ownership which had helped to make the land reform of the 1950s a success, had resulted in farms which were too small for productive use of labor in the 1970s. The Accelerated Rural Development Program which began in 1973 provided for local cooperatives which would enable farmers to jointly till adjacent land and to thereby increase the average amount of land cultivated by active farmers and encourage greater mechanization (Thorbecke, 1979, p. 192).

The policies of the 1970s and early 1980s appear to have been successful in keeping Taiwan's economic growth going at a healthy rate. Despite the recession in 1974 and a slowdown in the early 1980s, Taiwan averaged an economic growth rate of 8.8 percent per year from 1970 to 1975 and 10.6 percent per year from 1975 to 1980 (see Table 1.1). However, from 1980 to 1985, growth dropped to 6.4 percent per year, which was still very respectable by international standards. Taiwan also succeeded in broadening its exports and in increasing the number of countries in which they were sold. While textile products remained the largest single export, their share of total exports declined from 30 percent in 1971-75 to 21 percent in 1981-85 (Wu, 1985, p. 10). However in 1985, the proportion of exports going to Japan and the United States, the largest two trading partners, still exceeded 50 percent.

FACTORS INFLUENCING ECONOMIC DEVELOPMENT

While this review of Taiwan's economic development has focused on economic policies and their outcomes, there have been two other important factors which have contributed to the success of these policies. The first has been the continuing political stability and the confidence in the government which has enabled policies to be carried out. Gold credits the Nationalist party-state with providing sustained economic development through a series of crises. He states: "It did not just get the prices right, but it restructured society, channelled funds for investment, intervened directly in the economy, created a market system, devised indicative plans, determined the physical and psychological investment climate, and guided Taiwan's incorporation into the world capitalist system (Gold, 1986, p. 122). Pang

(1987) shows how a small group of dedicated leaders in the Kuomintang were successful in providing stable leadership because of their internal cohesion and their autonomy from the Taiwanese society.[6]

The second important factor in Taiwan's growth is the cohesiveness of the Chinese family. The importance of the family can be seen in the large number of family businesses, the reliance on family members for loans and other forms of support, the strong support which parents provide their children for higher education and other investments, such as migration, which will enhance their economic advancement, and the continuation of joint living arrangements after marriage. The characteristics of the Chinese family which differentiate it from families in other cultures have been described by Cohen (1976), Freedman (1966), Gallin (1966), Greenhalgh (1984), and Thornton, et al. (1987). These aspects of the Chinese family will be discussed in greater detail in the chapters to follow.

POPULATION GROWTH AND DEVELOPMENT

As was the case in many other developing countries, Taiwan experienced rapid population growth following World War II due to the combined effects of a reduced death rate and a postwar baby boom. As shown in Table 1.3, the crude death rate had declined to less than 10 per 1000 by the early 1950s and it continued to fall until it reached a level of about 5 per 1000 in the 1970s.[7] In contrast, the birth rate, which had been somewhat depressed during World War II, increased to 46 per 1000 between 1951 and 1955, higher than in any previously recorded period. While the higher death rates in earlier periods had partly offset the high birth rates and kept natural increase in a range around two percent per year, the combination of a record high birth rate and a low death rate resulted in a natural increase rate of 3.6 percent per year for the early 1950s. Although the birth rate declined in the 1950's to its prewar level of a little over 40 per 1000, natural increase remained above 3.0 percent for the entire takeoff period of Taiwan's development.

In the early 1960s, as Taiwan entered the labor-intensive period of development, the birth rate declined to 35.8 per 1000. In the following 10 years, the birth rate continued to decline, reaching a plateau around 24 per

----

[6] This autonomy, which was due to the fact that the initial leadership came from the mainland, is gradually disappearing as more and more Taiwanese are being elected and appointed to high government positions.

[7] The crude death rate in Taiwan was actually below that of many Western countries, although life expectancy was not as high. This was due to the fact that Taiwan had a very young age distribution so that there were fewer persons in the older ages where mortality was high.

TABLE 1.3

GROWTH OF POPULATION AND LABOR FORCE IN TAIWAN

Year	Year End Population (in 1000s)	Average Annual Growth Rate of Population	Average Crude Birth Rate /1000.	Average Crude Death Rate /1000	Average Annual Growth Rate of Labor Force
1921-25	3993	2.0	41.6	24.0	0.4
1926-30	4593	2.5	44.2	21.7	1.3
1931-35	5212	2.6	45.0	20.4	2.2
1936-40	5872	2.7	43.8	19.7	2.2
1941-45	6915	2.7	40.6*	17.6*	na
1946-50	7554	1.8	40.9	14.3	1.3
1951-55	9078	3.8	46.3	9.5	2.3
1956-60	10792	3.5	41.7	7.6	2.6
1961-65	12628	3.2	35.8	6.1	2.9
1966-70	14676	3.1	29.1	5.3	3.8
1970-75	16150	1.9	24.0	4.7	4.0
1975-80	17805	2.0	24.3	4.7	2.9
1980-85	19258	1.6	20.6	4.8	2.9

Sources: Data for 1921 to 1970 from Lee and Sun (1972). Data for 1970 to 1985 from Taiwan-Fukien Demographic Fact Book 1976 and 1985.

*Vital rates based on 1941-43 only. Labor force data not available.

1000 in the 1970s. When Taiwan entered the third stage of development in the mid-1970s, it had a rate of natural increase below 2.0 percent. In the period from 1980 to 1985, the birth rate fell further, bringing the natural increase rate down to about 1.6 percent per year.

Taiwan's active family planning program, which was initiated on an experimental basis in 1959 and expanded to cover most of Taiwan in 1964, is credited with accelerating the fertility decline during the labor-intensive stage of development (Freedman and Takeshita, 1969; Sun, 1977). There is considerable debate about the extent to which Taiwan's fertility decline is attributable to the family planning program (Freedman, et al, 1974; Hermalin, 1976) or due to the economic development and modernization of the country (Davis, 1967; Hauser, 1967). While avoiding that debate, it is possible to point to some of the effects of population growth on economic development and urbanization.

While the population growth rate was extremely high during the takeoff phase of Taiwan's development in the 1950s, the rate of growth of persons of labor force age was only about one half that amount, because the number of persons entering the labor force was related to the natural increase rate about 15 years earlier, which had been much lower (Mueller, 1977). This meant that the economy was not overwhelmed by large numbers seeking jobs. This timing was fortunate because the growth in employment was just barely sufficient to keep up with the growth of the labor force during this period.[8] The development of new industries did, however, require people to move to the sites of these industries which were usually in or near the larger cities and, as will be shown in chapter 2, there was rapid urbanization.

The rapid population growth of the 1950s probably had a negative effect on development because large families tended to save less than small families (Mueller, 1977). However, the timing of the population growth was fortuitous because it coincided with the period of maximum foreign assistance. Had it not been for this assistance, the takeoff of the economy might have been delayed until the dependency burden was less and the potential for domestic savings greater.[9]

The children of the postwar baby boom began to reach labor force ages in the mid 1960s and the growth in the labor force exceeded three percent

----

[8] According to Kuo (1983:53-63), the economy failed to absorb all newcomers during the 1950s and there was a small increase in unemployment.

[9] If one accepts the position that the main cause of fertility decline on the 1960s was the development which occurred in the 1950s, then it follows logically that without this development, fertility would have remained high in the 1960s and the dependency burden would have continued to have a negative effect on savings.

during this period (see Table 1.3). While the growth of the labor force tended to follow the natural increase rates about 15 years earlier, the growth in the labor force was further augmented by increasing participation of women. Liu (1983) shows that the proportion of women of labor force age who were in the labor force increased from 20 percent in 1956 to 25 percent in 1966 and 31 percent in 1970. This rapid growth in the labor force coincided with the expansion of labor-intensive industries in Taiwan and the availability of large numbers of young workers, who had at least six years of education and who were willing to work for low wages, helped make these industries competitive in an international market. According to the unemployment series based on the labor force surveys, the percentage of the work force who were unemployed dropped from about five percent in the early 1960s to less than two percent in the early 1970s (Kuo, 1983, pp. 332-333).

By the late 1970s, when Taiwan was in its third stage of development, the effect of the fertility decline of the 1960s began to have an effect on the number of persons reaching labor force age. In addition, many young persons continued their education into senior high school and college, thereby delaying their entry into the labor force. Initially these reductions in the growth of the labor force were offset by greater participation of women. Many industries which had had a policy of hiring only unmarried women, began to accept married women as the supply of unmarried women became insufficient to meet their labor demands. Liu (1983) shows a significant increase in the participation rate of married women between 1976 and 1980. Nevertheless, by 1985, there was widespread concern for labor shortages as the rate of growth of the labor force continued to fall, reflecting earlier declines in fertility.

No less significant than the changes in population growth rates were the changes in population composition which occurred between 1950 and 1985. While there had been serious efforts toward expanding education during the period of Japanese occupation, in 1946, about 55 percent of the population aged 6 and over was illiterate (Kuo, 1983, p. 123). By 1979, illiteracy had declined to only 11 percent and was limited mainly to persons who had grown up before the mid 1950s. In 1968, free education was extended to age 15 to include junior high school and most youths took advantage of this.[10] Since then there has been a steady increase in the proportion of young persons who continue on to high school and attend a college or university. In 1979, 52 percent of the youths aged 15 to 17 were attending senior high school and 28 percent of those aged 18 to 21 were attending a college or university (Kuo, 1983, p. 123). Increases in education played a significant role in enabling Taiwan to move from the labor-intensive phase of development to the industrial upgrading phase in

----

[10] In 1979, 91 percent of the youths aged 12 to 14 were attending junior high schools (Kuo, 1983:123).

the late 1970s.

CONCLUSION

In this chapter we have briefly reviewed Taiwan's economic development and the relationship between population growth and development. We have shown that Taiwan was fortunate both in experiencing rapid economic growth and in the timing of this growth so that the peak demand for labor during the expansion of labor-intensive, export oriented industries occurred when there was the most rapid growth in persons reaching labor force age. In the rest of this monograph we will show how these favorable trends in economic growth and population growth have affected the urbanization of Taiwan. While the growth of trade and industries and the high rate of population growth provided strong forces toward urban growth, the amount of urban growth and its distribution throughout Taiwan was not a simple function of these forces. Other factors such as the continued strength of the farm family as an economic unit and the development of extensive commuting networks and rural industrialization all contributed to make the rural to urban transition in Taiwan less painful than it might otherwise have been.

In the chapters to follow we will see how urbanization has responded to economic development and population change in Taiwan. In chapter 2, we will describe the growth of cities and the extent to which the growth has been balanced versus concentrated in and around one major city. Chapter 3 presents and tests an aggregate level model of growth among different urban areas over time. Chapter 4 provides an in depth look at Taipei and its periphery. In chapter 5, the relative roles of migration, commuting, and rural industrialization in the process of urbanization and development are investigated.

Chapters 6, 7 and 8 present results from a migration survey conducted in 1973. The timing of this survey places it in the middle of the period covered in the rest of this study and, as will be shown in the discussion of economic development, at an important stage in the increased industrialization to expand the export of manufactured goods. This was a time when the course of urbanization could have gone in many different directions. Chapter 6 tests a model of the individual determinants of migration and discusses the factors which determine who moves to the city and who remains behind. Chapter 7 investigates the economic consequences for the migrants both in comparison with those who remain behind at the place of origin and the natives at the place of destination. Finally, chapter 8 looks at the consequences of migration in terms of living conditions, social interaction, and "modern" attitudes.

The concluding chapter, chapter 9, summarizes the results and attempts to point out the problems which lie ahead in the further urbanization of Taiwan and some of the policies which may help to alter future trends in desirable directions. We shall also try to review the factors responsible for balanced urban growth in Taiwan and discuss the extent to which these same factors could apply to other less developed countries. Which aspects of Taiwan's urbanization are judged to be favorable and worth repeating in other settings? Which of the favorable aspects of Taiwan's urbanization can be repeated by other countries which are able to produce rapid economic growth? What aspects of Taiwan's favorable urbanization experience can be reproduced by countries with lower rates of economic growth? While we cannot provide definitive answers to all of these questions, we hope that this study will provide a better understanding of urbanization in Taiwan and its relationship to economic development and that this understanding will aid scholars and policy makers who are grappling with the problems of urbanization in other less developed countries.

2

An Historical Overview of Urban Growth

Compared with many developing countries, the growth of cities in Taiwan has been relatively smooth. While urbanization has kept pace with industrialization, the growth has been distributed over both time and space so that there have been few periods of very rapid growth and several cities have shared in the process. In this chapter we will review the historical growth of cities in Taiwan and the regional distribution of this growth. We will then show how urban growth has led to a system of cities which approximates the ideal rank-size distribution and avoids the excessive growth of a single primate city as has happened in several countries. In this process, intermediate cities have played an important role and we shall devote the last section to discussing the growth and employment conditions of these cities.

THE HISTORICAL GROWTH OF CITIES

When the Japanese gained control of Taiwan in 1895, it was a rural society with few settlements large enough to be considered to be cities. During their first three decades of colonial rule, Japan saw Taiwan mainly as a source of agricultural products which could be exported to Japan and most of the investment went into agricultural development and construction of transportation facilities necessary to get farm produce to ports. Consequently, urban growth was relatively slow.

The growth of Taiwan's cities can be traced from the early 1900s using published county and township data. The entire island is divided into municipalities and counties and these, in turn, are divided into districts and townships. With few exceptions, the boundaries of the townships and districts have remained unchanged over time, although some townships adjacent to municipalities have been added to these municipalities and some townships have been reclassified as municipalities. Because urban reclassification has proceeded in terms of whole townships or districts rather than the piecemeal annexation which occurs in much of the world, population data exist for each of these units and it is possible to trace their growth over time.

At the time of the 1930 census, seven local areas were given the rank of municipality (Barclay, 1954:103-113). These ranged in size from Taipei with a population of about 163,000 to Hsinchu with about 45,000. Two more areas were given the rank of city in 1932. By 1950, when the Chinese had regained control of Taiwan, there were 11 municipalities, five of which were designated as large cities and six as small cities (shih). The large cities were distinguished from the small cities both by their larger size and the fact that they were subdivided into districts (Ch'u), while the small cities were not. As townships grew in population and took on urban functions, they were reclassified as cities, although this reclassification was not automatic and often lagged the population growth by several years. Between 1950 and 1980, eleven areas were raised to the status of small cities and all of these had grown to over 100,000 before being reclassified. In the interim, all but one of the 11 municipalities of the 1950s had achieved populations over 100,000.[1] Thus the correspondence between official municipalities and places with populations over 100,000 was a close one at that time.

In 1982, Hsinchu and Chiayi were promoted from small cities to large cities which meant that they were no longer under the rule of county governments but were equivalent to counties in the administrative system. At the same time, county seats which were not already cities were classified as small cities regardless of size. By 1985, there were seven large cities and 20 small cities. Among the small cities, five had populations of less than 100,000.[2] This means that the official classification is no longer consistent with a fixed size rule for defining cities.

A better picture of urban growth can be obtained by using a fixed size rule such as all places with 100,000 or more population. This is possible because annual data are available from the household register for all townships in Taiwan allowing us to define which townships we consider to be cities and to follow them over time. The growth of places with populations of 100,000 and over is displayed in Table 2.1. In 1920 and

----

[1] The exception was Ilan which had a population of only 81,935 in 1980. However in 1980, one town, Yuanlin in Changhwa County, with population of 102,554, had not been reclassified as a city. Thus, the difference between the total population of places over 100,000 and the total population of officially designated cities was only about 20,000.

[2] The five new cities with populations under 100,000 in 1985 are Miaoli (84,994), Nantou (91,376), Touliu in Yunlin County (87,775), Hsinying in Tainan County (69,252) and Makung in Penghu County (55,177).

1930, Taipei was the only city with a population over 100,000. It had a growth rate of 3.5 percent per year during the 1920s which was faster than the 2.3 percent average annual growth rate for the total population, but still modest. If we were to assume that the rate of natural increase in Taipei was the same as the total growth rate of Taiwan, then natural increase accounted for about two-thirds of the total city growth.[3]

During the 1930s, Japan was preparing for war and expanded its goals for Taiwan to include processing of sugar and other food products, coal mining and manufacturing of aluminum, chemicals and some other products (Barclay, 1954, pp. 28-42). This lead to more rapid urban growth. By 1940, Tainan, Keelung and Kaohsiung joined Taipei as cities of over 100,000 and the proportion of the population living in cities of over 100,000 had more than doubled from about 5 percent to about 12 percent.[4]

Following the defeat of the Japanese in 1945, almost all of the Japanese who were living in Taiwan returned to Japan. The control of Taiwan was returned to the Chinese and some officials moved from the mainland of China to Taiwan immediately after the war. In 1949, when the communists took over the mainland, the government of the Republic of China and about 600,000 military men moved to Taiwan. While efforts were made to find rural land for some of these migrants, most of them settled in cities. Thus, while many of the government officials from the mainland filled positions vacated by the Japanese, there was substantial growth of cities during the 1940s. The number of cities over 100,000 increased to seven and the proportion of the population living in these cities exceeded 20 percent. Growth within constant boundaries averaged 5.6 percent which was more than double that of the total population. This was the only decade in which net migration appears to have exceeded urban natural increase as a source of growth. However, much of this migration was from outside Taiwan.

The 1950s and 1960s saw continued high rates of city growth, averaging over six percent per year when city expansion is included and about five percent when growth is calculated within constant boundaries.

As we have shown in chapter 1, the 1950s was a period of economic take-off when the overall population growth exceeded 3 percent per year, but the growth of the labor force aged population was considerably slower. While the birth rate declined in the 1960s resulting in a lower rate of

----

[3] The population growth of Taiwan was also affected by the immigration and emigration of Japanese and others. However the data presented by Barclay (1954, p. 13) show that net immigration accounted for only about 0.1 percent of annual growth in the 1920s.

[4] If one makes the comparison within constant boundaries, these four cities contained 8.9 percent of Taiwan's population in 1920 and 10.1 percent in 1930.

TABLE 2.1

GROWTH OF THE TOTAL POPULATION OF TAIWAN AND THE POPULATION IN CITIES OVER 100,000 FROM 1940 TO 1985

Year	Total Population (in 1000s)	Average Annual Growth in Previous Period	Number of Cities 100,000+	Population in Cities over 100,000	Average Annual Growth in Preceding Period: Overall (Percent)	Average Annual Growth in Preceding Period: Const. Bound (Percent)	Percent of Total Population in Cities
1920	3655		1	163			4.5
1930	4593	2.3	1	231	3.5	3.5	5.0
1940	5872	2.5	4	701	11.8	4.2	11.9
1950	7554	2.6	7	1560	8.3	5.6	20.7
1960	10792	3.6	9	2795	6.0	5.3	25.9
1970	14676	3.1	14	5282	6.6	4.7	36.0
1980	17805	2.0	22	8396	4.7	3.5	47.2
1985	19258	1.6	24	9668	2.9	2.4	50.2

Sources: 1920–1940: G. Barclay, Colonial Developmemt and Population in Taiwan (Princeton)
1950 and 1960: Household Registration Statistics of Taiwan 1959-61. Dept. of Civil Affairs, Taiwan Provincial Government, July 1962.
1970: Taiwan Demography Monthly 5:12 (December 1970).
1980: 1980 Taiwan-Fukien Demographic Fact Book.
1985: 1985 Taiwan-Fukien Demographic Fact Book.

natural increase, the growth of the labor force was greater due to the maturing of persons born during the baby boom. In both decades, this natural increase was augmented by a steady flow of rural to urban migrants so that the population living in cities over 100,000 more than tripled in 20 years. By 1970, there were 14 cities containing 36 percent of Taiwan's population.

During the 1970s, natural increase declined to about 2.0 percent per year, although migration to cities continued, so that by 1980, almost one half of the population of Taiwan lived in one of the 24 cities with a population over 100,000. As Table 2.1 shows, the rate of city growth slowed after 1980 and was only 0.8 percent higher than the growth rate for all of Taiwan. This was, however, sufficient to push the proportion living in cities over 50 percent. Much of the growth during this period was in peripheral areas of the larger cities or in satellite cities and not in core areas. The slowing of urban growth in the 1980s corresponds to the economic transition from an emphasis on labor intensive industries to greater emphasis on technology and services.

Throughout the twentieth century, the cities of Taiwan were distributed among the different regions in Taiwan in a system of cities. Figure 2.1 shows the location of all cities over 100,000 in 1985. In studying this distribution, it is helpful to know something about the physical geography of Taiwan. Although Taiwan is often thought of as a flat plain suitable for rice cultivation, only the West Central part of the island fits this description. About two-thirds of the area of Taiwan is mountainous with peaks as high as 13,064 feet (Hsieh, 1964). Along parts of the East region, these mountains extend to the sea, and in the rest of the East region, the width of flat land and low hills is only 10 to 15 miles from the coast to the high mountains. This means that this region has limited agricultural potential, although maximum use has been made of the land for growing pineapple and tree crops. The high mountains have also limited the access to the East region. Prior to 1960, when the cross-island highway was completed from Taichung to Hualien, persons wishing to travel to the East coast had to use roads at the extreme North or South of the island.

As a result of the geographical barriers separating the East Coast from the rest of the island, the growth of urban areas has primarily been restricted to the Western plains area and the low hills around Taipei in the North. In these areas, there appears to be a system of cities with Taipei and Kaohsiung representing the major growth poles in the North and South, and Taichung and Tainan, the third and fourth largest cities located nearer the center. Scattered around and between these four major cities are several smaller cities which represent sub-regional centers and manufacturing centers, the largest of which were Hsinchu and Chiayi. All of these cities in the Western part of Taiwan are linked by a network of highways and railroads which were initiated during the period of Japanese occupation and have been extended considerably since that time.

FIGURE 2.1

TAIWAN REGIONS AND MAJOR CITIES

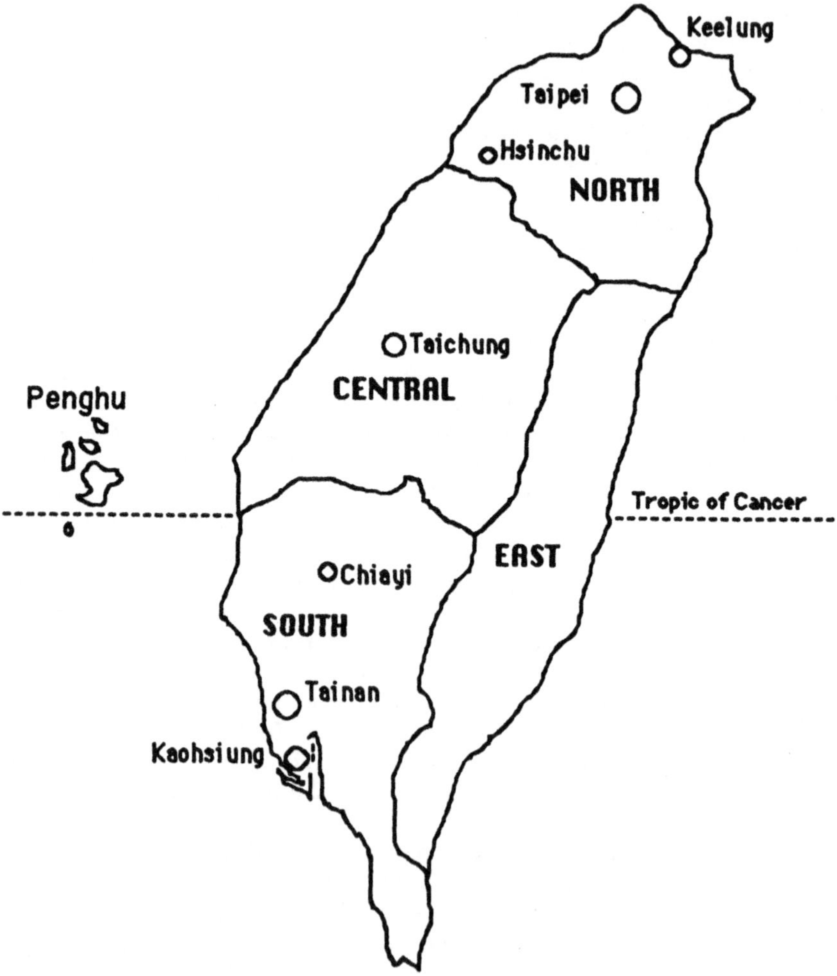

GROWTH OF LOCALITIES AND URBANIZED AREAS

There are two problems with the use of the official cities and townships to study urban growth in Taiwan. First, for some of the larger cities such as Taipei, Kaohsiung and Taichung, urbanization has spread beyond the city boundaries. While some adjustment for this spread has been made by occasionally adding adjacent townships to these cities, they continue to be underbounded.[5] For these cities, the use of constant boundaries understates urban growth. Secondly, when we try to look at the growth of smaller urban areas, we are confronted with the fact that most areas which are classified as urban townships contain an urban center surrounded by considerable rural area. Because rural population densities in the plains areas of Taiwan are high, a significant proportion of the population of these towns may be rural and thus the total population of the town is not a good indicator of the size of the urban center. Furthermore, some rural townships have developed significant urban centers but have not been reclassified as urban townships.

Definition of Localities and Urbanized Areas

In 1972, an attempt was made to develop a set of more precise urban-rural definition for Taiwan (Liu, 1975). The approach was to identify urban localities by classifying the small administrative areas of city wards (lis) and villages through the information provided by: (1) population statistics of wards and villages tabulated from the Population Register; (2) the urban characteristics of wards and villages gathered from a special survey; and (3) maps of administrative areas of wards and villages using a scale of 1:25,000.

In this new classification system, a locality is defined as one or more adjacent wards or villages which are within the same township or city and meet one or more of the following criteria:

1. The percentage of employed males in nonagricultural activities is 60 percent or more.

2. The ward or village has at least three of the following urban facilities: Kindergarten, primary school, junior or senior middle school, college or university, hospital, clinic, post office, cinema, recreation area, or public park.

----

[5] In 1968, six townships were added to Taipei. In 1979, one township was added to Kaohsiung City; and in 1982, one township was added to Hsinchu City. These were the only annexations between 1950 and 1985.

3. The ward or village has a population density of 2,000 or more registered residents per square kilometer.

4. The ward or village is the seat of township government.

5. The ward or village is surrounded completely or surrounded on three sides by qualifying wards or villages.

All localities with populations of 500,000 or more were designated as major cities. Those localities with a population of 100,000 or more registered inhabitants are defined as intermediate cities, and those with a population between 20,000 and 99,999 are defined as urban towns. Using these criteria, there were 15 cities and 52 urban towns in Taiwan in 1972, in contrast to the officially designated 16 cities and 71 urban townships.

The cities and urban towns determined by this new definition are still restricted to the official administrative boundaries of cities and townships. The new city and urban town definitions to some extent overcome the problem of overbounded cities. However, the problem of underbounded cities is not yet solved. To cope with this problem, statistical units called "urbanized areas" and "metropolitan areas" were established. An urbanized area is defined as a place which has a locality of 20,000 or more registered inhabitants and all the adjacent localities regardless of their size. A metropolitan area is defined as an administrative area which has an urbanized area of at least 100,000 or more registered inhabitants and the surrounding cities or townships with localities of 20,000 or more within their administrative boundaries.

As a result of combining adjacent localities across the administrative boundaries of cities and townships, 42 urbanized areas were defined in 1972. The metropolitan definition gives 9 metropolitan areas in Taiwan in 1972. The distribution and land areas of the urbanized areas and metropolitan areas are illustrated in Figure 2.2.

Obviously, the urbanized areas as set out by this definition are the more accurately measured units of urban centers since they are not restricted by administrative boundaries. Due to the inaccessibility of population data for the smallest administrative units (lis and villages) for earlier years, it is impossible to extend this precise measurement to ascertain long-run trends in the development of Taiwan's urban population. One of the best alternatives however, is to regroup the civil divisions of city precincts and urban and rural townships in accordance with this urban hierarchy of metropolitan and nonmetropolitan areas as established by the new set of definitions in 1972. The regrouped metropolitan areas, of course, contain some rural areas; the earlier the years in question in the study period, the larger the extent of this rurality. Taking into consideration the concentration of population in large metropolitan areas, our classification provides a better measure of urbanization than the use of the urban areas as designated at the beginning of the twentieth century.

FIGURE 2.2

URBANIZED AREAS AND METROPOLITAN AREAS OF TAIWAN REGION, 1972 AND 1980

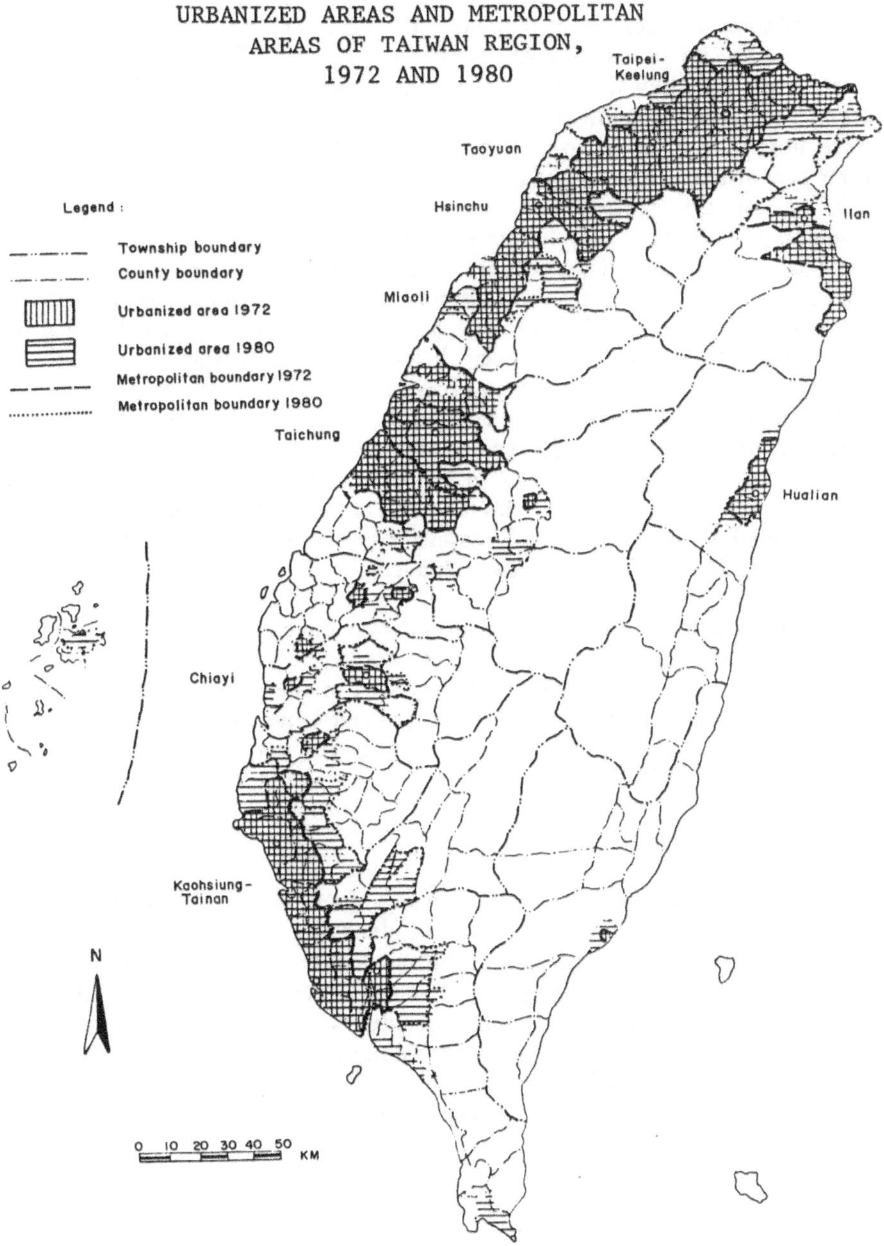

Levels of Urbanization

The levels of urbanization as measured by the cumulative percentages of population by locality of Taiwan and other selected countries around 1970 are shown in Table 2.2. It can be seen that Taiwan in 1972 was more urbanized than Canada in 1966 and New Zealand in 1971, when the urban population is defined as the population in localities with 500,000 or more inhabitants; otherwise, Taiwan was only slightly less urbanized on other size criteria than Canada, the U.S., Australia and New Zealand. The lack of strictly comparable data for other countries, especially for developing countries and countries which are similar to Taiwan with respect to land size and dependence on foreign markets such as Japan and England, limits generalization. However, judging from one aspect of economic development, Taiwan may be considered to be over-urbanized in the sense that at comparable levels of urbanization, it had a correspondingly greater proportion of its labor force engaged in agricultural employment (see the last line in Table 2.2). But if the hallmarks of "over-urbanization" are the excessive migration of the unemployed and underemployed rural population to the cities in advance or in excess of adequate expansion of urban employment opportunities, Taiwan does not belong to the "over-urbanized" category. An intensive microstudy on migration of Taiwan in the 1960s concluded:

> With a few exceptions, rural-to-urban migration in Taiwan can be viewed as a rational response to changing economic conditions. The growing rural population has adjusted to the pressure of population on limited resources by sending to major cities those members of the rural population who are best qualified for urban employment (Speare, 1974a).

Urbanization Trends

The trends of urbanization of Taiwan's population from 1930 to 1973 are presented in Figure 2.3.. These data show the distribution of the total population among various parts of metropolitan and nonmetropolitan areas over time. During the entire period of observation, the large metropolitan areas have been defined as the same six cities and their suburbs with fixed boundaries--Taipei-Keelung, Taichung-Changhwa, and Tainan-Kaohsiung. The small metropolitan areas consist of the cities of Taoyuan-Chungli, Hsinchu, Miaoli, Ilan-Lotung,[6] Chiayi and Hualien.

----

[6] Ilan-Lotung was included even though the population living in the urbanized areas of neither Ilan nor Lotung had attained a population of 100,000 in 1972.

TABLE 2.2

URBANIZATION IN TAIWAN AND OTHER SELECTED COUNTRIES AROUND 1970 (CUMULATIVE PERCENTAGE OF POPULATION)

Size of Locality	Taiwan 1972	Canada 1966	U.S. 1970	Australia 1971	New Zealand 1971
500,000+	34.3	30.0	42.0	57.9	22.7
100,000+	46.1	50.3	55.5	64.5	47.0
50,000+	50.0	53.0	57.3	67.0	51.4
25,000+	54.0	56.8	61.2	71.6	64.0
10,000+	56.8	62.2	66.1	75.8	67.5
5,000+	50.2	65.6	70.0	79.5	72.6
Proportion of labor force engaged in agricultural employment	33.0	7.4	4.3	7.2	12.5

Sources: Taiwan: Liu (1975).
Canada, U.S., and New Zealand: adapted from Campbell Gibson (1973:74).
Australia: U.N. (1972).

For the past half-century, there has been a steady trend toward a rise in the proportion of the total population residing in large metropolitan areas, increasing from 37.6 percent in 1930 to 50.6 percent in 1973 and 57.6 percent in 1985. The share of the population in small metropolitan areas tended to become stabilized (with small fluctuations) at a level of about 14 percent. Thus all, the gains in population in large metropolitan areas came at the expense of nonmetropolitan areas, including urban, rural and aboriginal towns.

The pattern of growth between and within metropolitan and nonmetropolitan areas has undergone significant changes. Before the second World War, large metropolitan areas and their central cities were growing steadily, and at a faster rate than small metropolitan and nonmetropolitan areas, while during the War the relative growth rates of the latter were gaining slightly at the cost of the former. Immediately after

FIGURE 2.3

PERCENT OF THE TOTAL POPULATION OF TAIWAN
METROPOLITAN AND NONMETROPOLITAN AREAS
1930-1985

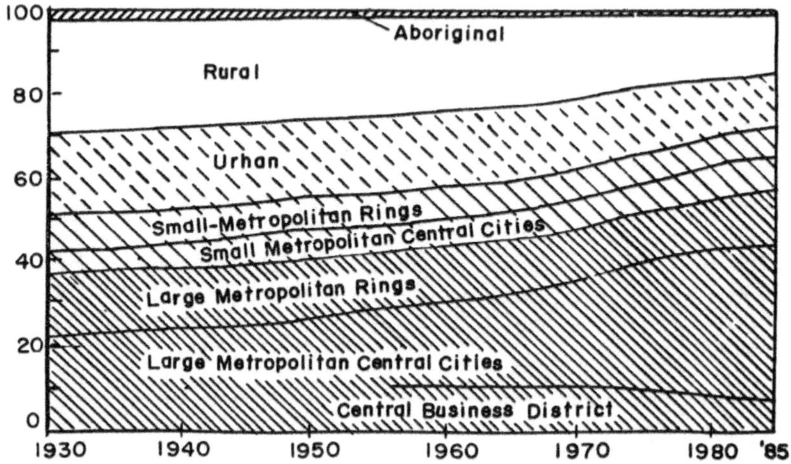

the War, large and small metropolitan areas gained population at the same rate from outside areas for a short period. Thereafter, the small metropolitan areas showed a sudden decline, and then fluctuated around the average rate of growth of Taiwan as a whole for the period from 1965 to 1968 and once again grew irregularly above the national level between 1968 and 1973. The large metropolitan areas, however, maintained their high relative growth rate until 1963, declined for a year, and then increased at an accelerated rate for the last decade.

The rates of growth of large metropolitan were not uniformly experienced in all their parts. Lack of detailed areal population data only permit us to trace these trends back to 1957. From the available statistics, and from scattered historical evidence from Taiwan and other countries, it might be safe to conclude that the process of urban concentration must first occur in the central business districts, which are all in the old part of the cities. The disadvantages of high central-area urban concentration growth as the development of modern transportation proceeds, and the relative growth rate of central business districts starts to slacken and then becomes negative (depopulation of the central business district). One may also speculate that the growth of large metropolitan rings started to become significant when the relative growth of the central business districts reached its plateau and when the central cities had experienced a relatively long period of high growth.

In sum, the pattern of growth trends of various parts of the metropolitan areas indicates that, although Taiwan has experienced a long process of urbanization, the major momentum occurred only after the start of sustained economic development in the 1950s. The direction and magnitude of the momentum were not uniform in all parts of the urban areas. The rings and the central cities of large metropolitan areas attracted the largest portion of rural migrants in the period of 1930 to 1960, and the rings have absorbed urban migrants from the central business districts since the late 1960s. The relative growth rates of large metropolitan rings and the central cities of large metropolitan areas thus may serve as more sensitive measurements of urbanization in Taiwan, and they will be used in connection with the changes in social, economic and demographic variables over time to test the causal model developed for the particular case of Taiwan.

THE GROWTH OF URBAN POPULATION BY SIZE OF PLACE

The number and population of all localities in Taiwan are presented in Table 2.3 by size and region. There were 67 cities and urban towns (i.e., localities of more than 20,000 residents) with a total of 7.7 million people in 1972. The urban population accounted for half of the population of Taiwan. The total area of the 67 cities and towns (1,914 sq. km.) was only 5.3 percent of the Taiwan surface, resulting in a high density of almost 4,000 persons per square kilometer. In 1980, the number of cities and towns became 90 with a total population of 11 million. This volume of registered residents raised its share in the total population of Taiwan from 50.1 percent to 62.1 percent in the eight year period. The total urban area was 2,628 square kilometers (7.4 percent of Taiwan) with an average density of 4,177 persons per square kilometer.

Between 1972 and 1980, there was an increase of 23 cities and towns in Taiwan. As shown in Table 2.3, eleven out of these are in the category of 50,000 to 100,000 population. The only two cities which entered the bracket of 500,000 or more people in the study period are Taichung and Tainan. In the meanwhile, the increment in the number of cities was 4 for the smallest size group and 3 for each of the two categories of 100,000 to 500,000 and 30,000 to 50,000.

The change in urban population from 7.7 million in 1972 to 11.0 million in 1980 indicates an increase of 43.3 percent in the eight years. The group of cities with 500,000 residents and over had the biggest increment (1.7 million) while the category of urban towns with populations between 50,000 and 100,000 enjoyed the highest growth rate (82 percent). As noted above, this category is also the one which had the largest number of new urban areas added to any category.

In terms of geographic distribution, Table 2.3 shows that the Central Region, gained the largest number of new cities and towns (10), followed by the Northern Region (7) and the Southern Region (5). The Eastern Region gained only one urban area, but this represented a 50 percent increase in the number of areas. Among the total increase in urban population, 45.2 percent belonged to the Northern Region. The Central Region which had

TABLE 2.3

NUMBER AND POPULATION OF CITIES AND TOWNS BY SIZE AND REGION, 1972 AND 1980

Population Size (in 000)	Taiwan No.	Taiwan Population in 1000s Persons	Taiwan %	North No.	North Population in 1000s Persons	North %	Center No.	Center Population in 1000s Persons	Center %	South No.	South Population in 1000s Persons	South %	East No.	East Population in 1000s Persons	East %
1972															
500+	2	2,793	18.3	1	1,891	35.0	0			1	902	18.0	0		
100-500	13	2,780	18.2	7	1,271	23.5	2	556	13.2	4	952	19.0	0		
50-100	12	824	5.4	6	405	7.5	2	155	3.7	2	106	2.1	2	158	25.0
30-50	22	827	5.4	8	316	5.8	10	361	8.5	4	150	3.0	0		
20-30	18	436	2.9	6	145	2.7	7	161	3.8	5	130	2.6	0		
All Cities	67	7,660	50.1	28	4,027	74.5	21	1,233	29.2	16	2,242	44.7	2	158	25.0
Total Pop.		15,289	100.0		5,407	100.0		4,227	100.0		5,022	100.0		632	100.0
1980															
500+	4	4,482	25.4	1	2,203	31.9	1	556	12.1	2	1,723	31.1	0		
100-500	16	3,436	19.4	10	2,440	35.4	2	270	5.9	3	622	11.2	1	103	16.2
50-100	23	1,502	8.5	10	665	9.6	8	520	11.3	4	243	4.4	1	74	11.6
30-50	25	1,017	5.8	11	463	6.7	9	370	8.0	5	184	3.3	0		
20-30	22	539	3.0	3	77	1.1	11	259	5.6	7	173	3.1	1	30	4.7
All Cities	90	10,976	62.1	35	5,848	84.7	31	1,975	42.9	21	2,946	53.2	3	206	32.5
Total Pop.															
Change 1972-1980*															
500+	2	1,689	60.5	0	312	16.5	1	556		1	821	91.0			
100-500	3	655	23.6	3	1,169	92.0	0	-286	-51.4	-1	-331	-34.7	1	103	
50-100	11	678	82.3	4	261	64.4	6	365	235.4	2	137	129.0	-1	-84	-53.3
30-50	3	190	22.3	3	147	46.4	-1	9	2.6	1	34	22.4	0		
20-30	4	103	23.7	-3	-68	-46.9	4	98	60.9	2	43	33.1	1	30	
All Cities	23	3,315	43.3	7	1,821	45.2	10	742	60.2	5	704	31.4	1	48	30.6
Total Pop.															

*Number refers to increased number of cities and population in category. Percentage refers to percentage increase in category.

the second biggest increase, registered the highest growth rate (60 percent) among the four regions. Both the Southern and Eastern Regions increased urban population by about 31 percent.

The data in Table 2.3 further reveal that the major increase of urban population in the Northern Region occurred in the category of intermediate cities ranging from 100,000 to 500,000 people (i.e., an increase of 1.2 million). Both the Central and Southern Regions share the same pattern of having most of their urban growth resulting from the entry of a city into the group of major cities with more than a half of a million population. Another pattern discovered for the two regions is that both have a significant growth in the category of minor cities of 50 to 100 thousand residents.

Table 2.4 shows the growth of urban population between 1972 and 1980 classified by the size of place in 1972. This particular way of classifying cities and towns by initial size is the "Individual City" method, which is different from the "Current Class" method employed for Table 2.3 (Davis, 1972:88-95). Table 2.4 does not contain any city or town which became qualified for a category during the study period and this means urban growth for a fixed set of areas. Consequently, the data in Table 2.4 show smaller amounts of growth and lower rates of increase than what were observed in Table 2.3 for most categories. This is especially the case in the group of major cities. Nevertheless, Table 2.4 provides direct information for assessing the growth of a specific group of cities by size. By this means, the examination of the relationship between urban growth and city size is facilitated. Data for the 26 localities which reached the minimum size of 20,000 residents required to be a city during the study period, are shown in parenthesis in Table 2.4.

According to Table 2.4, the urban population of the 64 cities and towns existing in 1972 increased by 32.7 percent in eight years.[7] In addition to natural increase and net migration, this increase includes the results from expansion of the existing urban areas and from annexation to them. However, it does not include the population of any locality which became qualified as a city through its own growth and expansion or through the integration of two or more smaller localities during the period. The two cities of more than a half of a million population in 1972 had a lower rate of growth (20 percent) than all the cities as a whole. The intermediate cities (100,000 to 500,000) had a slightly higher rate than the total, while

----

[7] As mentioned earlier and shown in Table 2.3, the number of cities and urban towns was 67 in 1972. Two of them were within the administrative boundaries of a township (Nantou) and became a city in 1980. Hsiaokang was classified as an urban town in 1972, but it was annexed to Kaohsiung City in 1979. Taya in Taichung County was marginally qualified for an urban town in 1972, but it was not classified as an urban town in 1980.

TABLE 2.4

GROWTH OF URBAN POPULATION BETWEEN 1972 AND 1980 BY REGION AND SIZE OF PLACE IN 1972

Region and Size of Cities in 1972	No.	Population in 1972	Population in 1980	Growth Persons	Percent
Taiwan	64*	7,639	10,135	2,496	32.7
500+	2	2,828	3,397	570	20.1
100-500	13	2,780	3,715	934	33.6
50-100	12	824	1,263	439	53.3
30-50	22	841	1,208	367	43.6
20-30	15	366	552	186	51.0
(<20)	(26)	(402)	(841)	(439)	(109.2)
North	28	4,027	5,583	1,556	38.6
500+	1	1,891	2,203	312	16.5
100-500	7	1,271	1,858	586	46.1
50-100	6	405	769	364	89.9
30-50	8	316	460	144	45.5
20-30	6	145	294	149	103.4
(<20)	(7)	(121)	(265)	(144)	(119.4)
Center	19	1,212	1,613	401	33.1
100-500	2	556	706	150	27.0
50-100	2	155	195	40	25.9
30-50	11	410	586	176	42.9
20-30	4	91	126	35	38.1
(<20)	(12)	(186)	(363)	(177)	(95.2)
South	15	2,242	2,763	521	23.2
500+	1	937	1,195	258	27.5
100-500	4	953	1,151	198	20.8
50-100	2	106	123	16	15.4
30-50	3	115	162	47	40.6
20-30	5	130	132	2	1.8
(<20)	(6)	(78)	(183)	(105)	(134.1)
East					
50-100	2	158	176	18	11.7
(<20)	(1)	(17)	(30)	(13)	(76.8)

*Three out of the original 67 were annexed during the study period.

Note: Figures in parentheses are for reference only. They are the data of the 26 localities which reached a population of 20,000 or more during the study period.

small cities grew much faster (ranging from 44 to 53 percent). For the 26 newly added areas, the growth rate was 109 percent. The data show that the smaller the size of a place is the faster it grows. This observation indicates an obvious suburbanization process occurring in at least some parts of Taiwan.

The breakdowns by region in Tables 2.4 further suggest that the Northern Region is unusual in many aspects. It had the highest growth rate of 38.6 percent among the four regions. The rates of increase of urban population in the intermediate, small and new cities were especially pronounced (about 90 to 100 percent). The Central Region had a growth rate which equals that of the whole of Taiwan (33 percent), followed by the 23 percent of the Southern Region and the 12 percent for the Eastern Region. The data of the Central and Southern Regions fail to depict a clear picture of suburbanization as observed for the Northern Region.

The difference between the results derived by the "Current Class" method (Table 2.3) and those by the "Individual City" method (Table 2.4) represents the effect made by the new cities (of a current class) on the population growth (of that class). For example, the growth of the Northern Region was mainly due to the population increase in the existing intermediate cities and urban towns. In contrast, the addition of the new areas was the major source of urban growth in the Central Region. This is where the land area of cities increased by 63 percent.

A comparison of the lists of cities and towns classified by size for 1972 and 1980 shows, in addition, that 13 areas in the Northern Region had their positions promoted from a lower to a higher class during the eight years. In the meantime, the region had only seven localities which became qualified as urban towns by passing the 20,000 minimum size. For the Central Region, there were 11 areas added to the list and nine which moved upward in the size-categories. Eight of the 11 new areas entered the class of the smallest cities (20,000 to 30,000 residents). There were six new areas and five promotions in the Southern Region. The Eastern Region had only one new town and one promotion.

THE RANK-SIZE DISTRIBUTION OF CITIES

Several writers on the relationship between urbanization and economic development have focused on the size distribution of cities (Zipf, 1941; Berry, 1961; Vapnarsky, 1969; Smith, 1985). These writers have argued that if development is spread throughout the country, that the natural distribution of cities fits a rank-size rule (Browning and Gibbs, 1961). According to this rule, the expected size of each city should be equal to the size of the largest city divided by the rank of the city or:

$P_r = P_1 / r$ where P_r = the population of the rth ranked city
P_1 = population of largest city
r = rank

distribution because the logarithm of P_r is proportional to negative r.

It has been observed that many less developed countries deviate significantly from this rule because the largest city is much larger than the next largest (Ginsberg, 1955; Linsky, 1969; Mehta, 1964). In such cases the largest city is referred to as a "primate city" and a prime example is Bangkok which in 1980 was 27 times the size of Chiangmai, the second largest city in Thailand.

More recently, Smith (1985) has argued that primacy and the rank-size distribution of cities are somewhat different in that primacy refers to the largest city while the rank-size rule applies to the entire distribution of cities. The hypothesis that a log-normal urban system without primacy will be found where "the potentially dominant class is the same at each level of the urban system and that class finds its interests served by open or competitive forms of labor organization" (Smith, 1985:131). However, if the politically dominant class is different between the highest level and lower levels, a primate city is likely to emerge in an urban system which otherwise has a log-normal distribution. Other combinations are possible when there are closed or noncompetitive forms of labor organizations.

Supposing the structure of size distribution of urban places in a country is of the "rank-size rule" type, Suzuki (1980) found that the structure of size distribution of urban places in any region of the country will also be of the "rank-size rule" type if certain conditions were met. This special characteristic of urban population distribution is referred to as the decomposability of the rank-size rule. In Japan, the rank-size rule was tested separately for nine regions with data of 1960, 1965, 1970, and 1975 and found to meet the conditions for decomposability. (Suzuki and Kuroda, 1982).

Table 2.5 presents the results for Taiwan, using the 1972 and 1980 data. The structure of size distribution of cities in the whole of Taiwan confirms to the rank-size rule for both years. The estimated values for the slope of the regression line are rather close to negative unity. With the only exception of the Eastern Region, the regional structures of city distributions are all of the "rank-size rule" type. The Southern Region has the highest degree of fit, followed by the Northern and Central Regions. The Eastern Region has too limited a number of cities to permit a detailed analysis. It is obvious that the decomposability of the rank-size rule was not uniformly followed in Taiwan. Nevertheless, the data clearly suggest that the structure of size distribution of cities in Taiwan is rather homogeneous from region to region, at least among the three major parts on the west of the country.

Having shown that the cities in Taiwan generally fit the rank-size or log normal distribution, we shall now look for evidence of primacy. In measuring primacy, we have used the four city index suggested by Browning and Gibbs (1961:440). This index is simply the ratio of the population of the largest city to the sum of the populations of the next three

TABLE 2.5

ESTIMATES OF RANK-SIZE RULE PARAMETERS FOR REGIONS OF TAIWAN, 1972 AND 1980

Year	Estimates	Taiwan	North	Center	South	East
1972						
	a'	1.05	1.18	0.85	1.41	-
	b'	6.23	6.05	5.39	5.99	-
	R^2	0.99	0.97	0.93	0.98	-
	n	67	28	21	16	2
1980						
	a'	1.03	1.12	0.81	1.37	-
	b'	6.37	6.20	5.53	6.06	-
	R^2	0.99	0.96	0.96	0.98	-
	n	90	35	31	21	3

Note: a' and b' are the two parameters in $\log P_s = -a' \log R_s + b'$, where P_s is the population size of a city in the $_s$th region. n denotes the number of cities.

cities. If the cities followed the rank size rule, this ratio would be 1/(1/2 + 1/3 + 1/4) or approximately .92. However, if there is primacy, this ratio will be considerably greater than 1.0.

According to the four city indices shown in Table 2.6, Taiwan has had relatively little primacy during the twentieth century. The index, which was about 1.0 in 1950, declined with development to .81 in 1960 and has since increased to a figure very close to the ideal based on the rank size rule.[8]

---

[8] The primacy indices for Taiwan were calculated using official city boundaries. Since urban growth has expanded well beyond the boundaries of some of these cities, some might argue that it would be better to use metropolitan areas as the units for this calculation. When this was done for 1980, the index of primacy was .74 which indicates even less primacy than the index based on official cities.

TABLE 2.6

FOUR CITY INDEX OF PRIMACY FOR TAIWAN AND OTHER ASIAN COUNTRIES

Taiwan	1920	1.01
	1940	.88
	1960	.81
	1980	.93
Bangladesh	1981	1.58
China	1982	.43
Indonesia	1980	1.34
India	1981	.72
Japan	1984	1.19
South Korea	1980	1.43
Philippines	1984	4.54
Thailand	1980	12.04

Sources: Taiwan 1920-1940 from Barclay (1954:116-119) 1960-1980 Taiwan Demographic Fact Book.

Other countries from United Nations, Demographic Year Book 1985, New York, 1987. Based on cities except for Bangladesh and Philippines where urban agglomerations were used.

Only a few other Asian countries approach this ideal. The two largest countries, mainland China and India have competing major cities which give them values considerably below the ideal. Japan, the most developed Asian country, comes closest to the ideal with a primacy index of 1.19. South Korea, Bangladesh and Indonesia show moderate amounts of primacy with the largest city being about 50 percent above the expected size. Thailand and the Philippines have primate cities which are several times the expected sizes.[9] Thus in comparison with other Asian countries,

---

[9] The computation of primacy is affected by whether the city proper or the larger metropolitan area is used for computation. In the Philippines, Manila is clearly underbounded and the urban agglomeration, as defined

Taiwan appears to come closer to the ideal level of primacy than any other country. One reason for this has been the role played by intermediate cities in the system of cities.

THE GROWTH OF INTERMEDIATE CITIES

Intermediate or secondary cities are typically defind as those ranging in size from 100,000 to about 500,000 (Rondinelli, 1983). Many intermediate cities are regional metropolises which play a key function in the national system of cities between the national cities and smaller cities (Meyer, 1984). Others are industrial cities which are important centers of manufacturing but which are not as important in the coordination of regional economic activities.

In this section, we will focus on the 17 cities which were officially designated as county level cities in 1980. These correspond fairly closely to the list of places with populations of 100,000 to 500,000 in 1980.[10] The growth of this set of cities is shown in Table 2.7. Between 1960 and 1985, the population living in these 17 cities increased three-fold from 1.3 million to about 3.9 million. In contrast, the five large cities increased by a factor of 2.25.

The growth of the intermediate cities was about the same as that of the large cities during the 1960s when natural increase was high. With the decline in natural increase in the 1970s, the growth of the large cities decreased to only 2.6 percent per year while the intermediate cities grew at almost twice that rate. In the early 1980s, growth rates for both large and intermediate cities declined, although the intermediate cities continued to grow more rapidly than the small cities.

The growth of the intermediate cities, however, was not entirely uniform. The old cities, those which had been designated as county level cities by 1960, grew much more slowly than the newer cities. During the 1980s, the old cities had an average annual growth rate of only 1.4 percent which was considerably below the national rate of natural increase (1.94 percent). Some of these cities such as Hsinchu and Chiayi were older manufacturing centers which had not attracted the newer growth

----

by the United Nations, was used.

[10] All of the county level cities had populations between 100,000 and 500,000 except for Ilan City which had only 81,935. However, one urban township Yuanlin, with a population of 102,554 was not included, so that the net difference was only about 20,600. In addition, Keelung City was classified with the large cities because it had historically been a provincial level city although its population was only 344,867. The other four large cities had populations in excess of 500,000 in 1980.

TABLE 2.7

POPULATION AND GROWTH - LARGE AND INTERMEDIATE CITIES
(Based on official Cities in 1980)

	1960	1970	1980	1985
	Population in 1000s			
Large Cities	2435	3845	4945	5477
Intermediate Cities	1306	2093	3430	3923
Old cities	671	915	1046	1149
New cities				
around Taipei	314	664	1541	1814
other new cities	320	514	842	960
Total all Cities	3701	5938	8375	9400
	Average Annual Growth in Preceding Period (Percent)			
Large Cities		4.7	2.6	2.1
Intermediate Cities		4.8	5.1	2.7
Old cities		3.2	1.4	1.9
New cities				
around Taipei		7.8	8.8	3.3
other new cities		4.8	5.1	2.6
Total all Cities		4.7	3.5	2.9

Note: Large cities are Taipei, Kaohsiung, Taichung, Tainan, and Keelung. All except Keelung exceeded 500,000 in 1980. The old intermediate cities include all of those classified as cities by 1960. These include Hsinchu, Chiayi, Pingtung, Changhwa, Ilan and Hualien. The new cities around Taipei are Sanchung, Panchiao, Chungho, Yungho, Hsinchuang and Hsintien. The other new cities are Fengshan, Taoyuan, Chungli, Fengyuan and Taitung.

industries. Others such as Pingtung, Hualien and Ilan were small regional centers in regions which did not attract much new economic activity during that period.

The new intermediate cities fall into two categories. Six of these cities are located on the periphery of Taipei. While their population have grown significantly since 1960, they are more than residential suburbs. Much of the new industrial development has taken place in these cities. Between 1960 and 1980, the population of these cities increased nearly six fold.

The other five new cities are scattered throughout Taiwan. Two of these, Fengyuan and Fengshan are satellite to other large cities (Taichung and Kaohsiung), while two others are manufacturing centers (Taoyuan and Chungli). The other city, Taitung, is a regional center on the East Coast. As a group, these other new cities grew at rates about equal to the average for all intermediate cities--faster than the old cities, but considerably slower than the cities surrounding Taipei.

Data on the distribution of employment by level of urbanization (see Table 2.8) show that the intermediate cities were roughly similar to the large cities in composition. The major difference is that the intermediate cities have a higher proportion employed in manufacturing and a lower proportion in transportation, and services. This is understandable since two of the large cities are the main ports of Taiwan. The intermediate cities also have a slightly higher proportion employed in agriculture indicating that some may have still been slightly over bounded in 1978.

In contrast to the cities, both the urban and rural townships had significant proportions of their work force employed in agriculture and much lower proportions in commerce. However, both urban and rural townships had significant employment in manufacturing. In urban townships, over one-third of the female workers were in manufacturing while over one-quarter of the women in rural townships were employed in manufacturing. Since these data are based on place of residence, some of these workers may commute to larger manufacturing centers. But there was also a considerable amount of manufacturing activity in smaller towns.

The intermediate cities compared favorably with the large cities in terms of employment and utilization of labor. Data on labor utilization for different areas of Taiwan were collected by a special survey in 1979. This survey was based on the labor utilization approach, devised by Hauser (1974). In contrast to the labor force approach which places emphasis on the employed and the unemployed, the labor utilization approach emphasizes the "adequately utilized." The inadequately utilized comprises the unemployed and the underemployed. The underemployed includes those who are inadequately utilized in terms of short hours of work, low income and mismatch between occupation and education. The priority is first given to hours of work, second to mismatch, and level of income comes last. The most controversial problem in this scheme is the choice of criteria for the cut-off points for determining underemployment. After three experimental surveys, the 1979 survey adopted the following set of criteria:

TABLE 2.8

PERCENTAGE DISTRIBUTION OF EMPLOYED POPULATION BY INDUSTRY
1978

	Large Cities	Inter-mediate Cities	Urban Town-ships	Rural Town-ships
	Males			
Agriculture	7.9	12.2	34.0	50.6
Manufacturing	31.8	35.2	25.7	19.6
Commerce	16.7	14.0	8.4	4.7
Transportation	10.6	7.5	5.1	3.7
Services	32.8	31.1	26.6	21.4
Total	100.0	100.0	100.0	100.0
	Females			
Agriculture	8.7	10.9	34.5	56.5
Manufacturing	33.5	43.7	35.6	26.4
Commerce	23.2	18.6	10.2	5.5
Transportation	4.0	2.4	1.7	0.9
Services	30.5	24.4	18.1	10.6
Total	100.0	100.0	100.0	100.0

(1) Hours of Work. Those who during the reference week worked less than 36 hours but who wanted more work are considered inadequately utilized by hours of work.

(2) Education Mismatch. A normal range of educational attainment was defined for each occupation and individuals who exceeded the maximum education in the range were considered mismatched.

(3) Level of Income. Individual workers were classified by educational attainment and employment status for each sex (36 categories). Those whose income was less than 50 percent of the average income of the category were considered inadequately utilized by level of income.

TABLE 2.9

LABOR UTILIZATION FOR THE INTERMEDIATE CITIES AND OTHER LOCALITIES OF TAIWAN, 1978

	Both Sexes		
	Large Cities	Intermediate Cities	Remainder
Adequately Utilized	89.2	88.1	85.2
Inadequately Utilized	10.8	11.9	14.8
by Unemployment	1.5	1.6	0.9
by Hours of Work	0.7	1.4	2.7
by Level of Income	2.0	3.7	7.8
by Education mismatch	6.6	5.2	3.4

Source: Special tabulation from May 1979 survey on labor utilization. Estimates of total population based on a sample of approximately 48,000.

A summary picture of labor utilization for different size of cities in Taiwan is given in Table 2.9. For the large cities, 89.2 percent of labor force was found to be adequately utilized while 10.8 percent was found to be inadequately utilized. The percentage of adequacy for the intermediate cities was only about one percentage point lower than the large cities, but the gap widened to about 3 percentage points lower between the intermediate cities and the remainder of Taiwan. The bulk of the underutilized in the cities of all size is education mismatch, while in the remainder of Taiwan most underutilization is by level of income. Inadequate utilization by level of income tends to be negatively associated with urbanization while that by educational mismatch tends to be positively associated with urbanization. Unemployment and inadequate utilization by hours of work were found to be less important. Unemployment showed a positive relationship to urbanization, while hours showed a negative relation to urbanization. The extent and pattern of labor utilization and their relation to urbanization were found to be quite similar for males and females.

These findings again support the argument that rural to urban migration of labor, either to the large cities or to the intermediate cities, had increased the adequacy of labor utilization through raising incomes in spite of a small increase in inadequate utilization by education mismatch.

An examination of the differentials in the incidence of labor utilization classified by demographic and socioeconomic characteristics for various size of cities was undertaken by Liu (1979). He found that in the aggregate there were no significant differences in patterns of male and female labor underutilization for various sizes of cities. When the utilization rates were disaggregated by age groups, variations in pattern among sexes appeared. Very high rates of inadequate utilization by level of income characterize the youngest male age group of 15-19, especially outside cities. The income inadequacies gave way to underutilization by education mismatch in the next age group of 20-24 with the highest rate of mismatch in the large cities. For males, as age advances the rates of adequate utilization increase to a peak in accordance to the degree of urbanization; they decrease steadily thereafter. In the large cities, the peak is 95 percent occurring in the age group 45-49, while in the intermediate cities, it is 96 percent in age group 40-44 and in the remainder it is 91 percent in the age group 30-34. The bulk of the inadequacy beyond age group 20-24 was found to be mismatches for most age groups, but inadequate incomes at old age groups.

In contrast, the rate of adequate utilization for females in the large cities is found to be much higher than for males in young age groups of 15-19, but it decreases sharply until about age 30, rises to a plateau around age 40 to 60 and then declines in advanced ages. The age patterns of adequate utilization for females in the intermediate cities and the remainder of the country are quite similar to the large cities but the amplitude of the fluctuation narrows with the decreasing degree of urbanization. The level of the rate of adequacy are also consistently negatively associated with urbanization.

There were several interesting aspects to the inadequate utilization of labor by level of educational attainment for various localities. The most significant features are: (1) The rates of inadequate utilization of both sexes are highly and positively related to the level of education, with an definite jump for the junior college degree holders. (2) These rates are negatively related to urbanization within each level of education. (3) The negative relation between underutilization and urbanization is significantly stronger for males than for females.

For males underutilization is most severe in the agricultural sector while, for females, underutilization is greatest in the service sector. Inadequate income and mismatch are the main sources of the underutilization. Lower rates of underutilization for both sexes characterize the manufacturing sector in all type of localities.

CONCLUSION

Despite rapid industrialization since 1950, Taiwan's urbanization has proceeded at a fairly steady pace which has avoided many of the problems associated with rapid and uneven urban growth. During most of the period, natural increase accounted for more than one half of the total growth.

Urbanization has been spread throughout a system of cities encompassing the entire island. The city size distribution approximates the ideal rank-size distribution and there is less evidence of primacy than in most other Asian nations. Of particular importance in the development of this system of cities has been the role played by the intermediate sized cities. In recent years, these cities have grown more rapidly than the large cities.

In later chapters we shall examine the process of migration to cities and the fate of migrants in cities. Data on labor utilization indicate that unemployment is very low in both the large cities and the intermediate cities and that the underutilization of labor as measured by insufficient hours worked, low income and educational mismatch was lower in the cities than in the rest of Taiwan. These data suggest that there are few migrants in cities who are having difficulties finding adequate employment.

On the whole, urbanization patterns in Taiwan conform closely to the urban-industrial growth experienced in the advanced western nations. Taiwan obviously is one of the few special cases of development and urbanization outside the western cultures. In the past, the driving force behind economic development and urbanization has been the growth of manufacturing to provide for import substitution and for exporting to the world market. However, the labor intensive industries in Taiwan are facing increasing competition from lower wage countries and further industrial growth is likely to depend on the development of capital and technology intensive industries.

3

The Relationship Between Urbanization and Development

One of the striking phenomena in many developing countries is the rapid rate of urbanization amidst a high level of urban unemployment. The theoretical and empirical relationships established on the experiences of the "advanced" industrial countries seem to provide inadequate explanations of this peculiar phenomenon. The Harris-Todaro model showed how conventional economic tools could be used to tackle this problem (Harris and Todaro, 1970). This model is a typical closed neoclassical two-sector growth model incorporating urban expected wage rate as a special variable. Empirical studies on African and Latin American countries, where urban unemployment problems are most serious, lend significant support for this reformulation (Beals et al. 1967; Sahoda, 1968; Annable, 1972; Ominde and Ejiogo, 1972; Berry, 1975). The extension of the Harris-Todaro model by Cordon and Findley (1975) to allow for capital mobility between sectors gives it considerably more research and policy implications. This direction of the development of the model is encouraging and promising, but the approach adopted is limited to a comparatively static situation and hence relies on the existence of a stable equilibrium in the long run.

The relationship between urbanization and urban unemployment encompasses a whole range of socioeconomic variables, and, as many sociologists, economists and demographers see it, involves circular causation in a cumulative process. This type of relationship is described by Jaffe and Stewart (1951) in the following statement:

> As commercial and manufacturing activities greatly expand, it becomes necessary for a much larger proportion of the population to assemble in large aggregations, in cities, for only in that way can these activities be carried on efficiently. The large-scale growth of urban centers, in turn, is one of the elements that affect the rate of population growth (p. 423).

What Jaffe and Stewart illustrate is a "virtuous" circle based on the experiences of the industrial countries. They assume that at the beginning of modern economic growth, technological progress first induced a decline in mortality through its effects on increasing food supply, improvements in means of transportation and in the field of public health; and thus causes an increase in population growth. Consequently, a rise of modern industry and urbanization followed. Large-scale urbanization then facilitated the decline of the birth rate because the close proximity of people permitted the dissemination of new values and motivations in favor of a small family and high mass consumption. Increased consumption (or demand) that came from increases in population size or increases in per capita consumption, or both, in turn, channelled the rate and direction of technological progress into further expansion of industrialization and urbanization. This is obviously a cumulative upward movement and has been called a "virtuous circle". This process can also be reversed and turned into a cumulative downward movement, i.e., the familiar "vicious circle" if the relative strength and timing of the determinants in Jaffe and Stewart's circular causation is disturbed, as Myrdal (1968) pointed out. What happened in most developing countries is that the sudden injection of cheap public health and medical innovations after the Second World War caused a sudden decline in mortality and then rapid population increase, while leaving food production and other conditions unchanged. The pressure of increased population on land with little available agricultural surplus cut short the developmental process of industrialization and produced a larger proportion of population living in urban places through both urban natural increase and the flow of rural migrants expelled from the land. The increases in urban population were largely absorbed in the traditional sectors without the need for new supporting employment, services and facilities. Contrary to the preceding situation, where urban industrialization led to urbanization, this new demographic trend meant that urbanization (or overurbanization) tended to stifle industrialization and stimulate further population growth (Mabogunje, 1972).

Taiwan, like most developing countries, experienced a rapid population growth following World War II due to a sudden decline in mortality resulting from modern cheap public health innovations and persistently high fertility immediately after the war. The population pressure on land in the already densely populated island was extremely severe. Taiwan has, however, managed to avoid the vicious circle and to push towards an upward spiraling development. The factors and policy measures which account for this success were discussed in chapter 1. The purpose of this chapter is to suggest that the undesired relationship between urbanization and socioeconomic development occurring in many developing countries could be turned into a desirable one by proper policy measures such as those implemented in Taiwan.

The following section presents a general causal model for both temporary and spatial analysis of the relationship between urbanization and socioeconomic development. Estimates of the variables and path coefficients included in the theoretical framework follow in the next section. The last section summarizes the basic findings and limitations of the model.

SPECIFICATION OF THE MODEL

In longitudinal and cross-sectional studies, the level or rate of urbanization is generally taken as a function of a variety of social, economic and demographic variables. However, the use of ordinary regression analysis to tackle a multivariable system with many high correlations among variables like the phenomenon of urbanization can not always provide a satisfactory way out of the puzzle. A good example is the so called "strictly empirical study" conducted by Leo Schnore (1961). In the study he calculated rank correlations between urbanization and ten indicators of modernization for seventy-five countries around 1950-1955, and found high correlations for almost all of the pairs. This result, together with the findings of previous similar studies lead Schnore to conclude:

> In point of fact, however, the proximate goal to be sought is a mutual interchange between the representatives of these two approaches; ideally, empirical data of the type considered here will aid the theorist in successively constructing models ever closer to the empirical world, while his abstract analytical models will sensitize the empiricist to new types of relevant information (p. 244).

Without a mutual interchange between these two, he seems to imply, "the theorist is unable to enter the closed system and the planner has no real chance of breaking the vicious circle" (Schnore, 1961, p. 243). True, the cooperation between empiricists and theorists is inevitable, but instead of pursuing their respective extreme goals it may be more appropriate to compromise by first specifying a logically consistent theoretical model, even one which is not so rigorous as might be desirable, and then to test this model with the available data. Our study attempts to integrate the urbanization experiences of Taiwan in recent decades into a theoretical model as a step in this direction. In doing this we have to identify both the primary-variables, i.e., those variables influencing the level or rate of urbanization directly, and also those influencing the level or rate of urbanization indirectly through their influence on primary variables. In Figure 3.1 such a model is constructed for Taiwan.

The most immediate, and probably the most important explanatory variable is industrialization. It is well established, both theoretically and empirically, that the distribution of population is highly dependent upon the changes in economic opportunities (Kuznets, 1957). Persons engaged in agriculture are extensive users of land and thus tend to spread out thinly over arable areas, while those employed in manufacturing and service industries are intensive users of space, and hence tend to concentrate in a

FIGURE 3.1

PATH DIAGRAM OF CAUSAL MODEL OF URBANIZATION

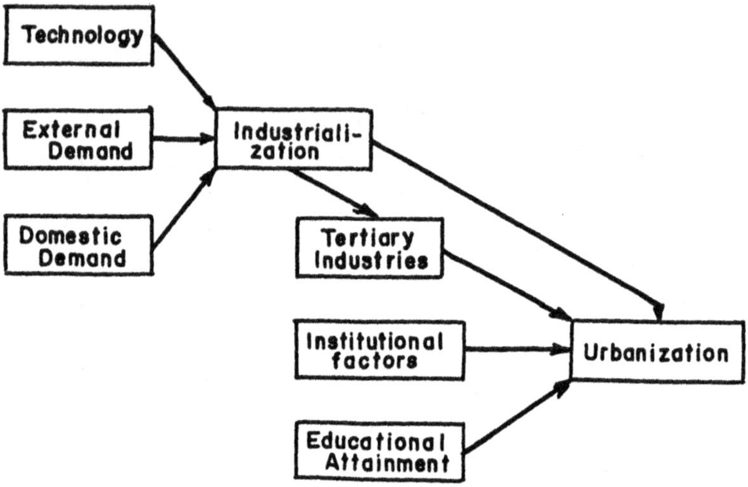

few localities. As an economy is in transition from an agricultural to an industrial economic base, the massing of population into the existing or newly formed urban centers goes hand in hand with industrialization.

Taiwan has been densely populated since the 1940s. By 1960, the overall density was 300 persons per square kilometer, and the average density of the five largest cities was 3,430 per square kilometer. The expansion of industries in urban centers was severely restricted by the scarcity of land. There were several advantages of locating industries in the countryside: relatively lower land prices, an inexpensive and convenient transportation network; accessibility to the surplus rural labor force; and industrial sites and infrastructure provided by the government at relatively low prices or for free. The primary effect of the dispersed industrialization was a substantial reduction of the flow of industrial workers from rural to urban agglomerations. But the services needed to support the expanding manufacturing industries, to a large extent, could grow in large cities without the limitation of land scarcity, for they were able to use space more economically and efficiently. The comparison of the growth rates of industrial and service employment in large cities and in the rest of Taiwan is shown in Table 3.1.

Based on these facts, it may be argued that the total effects of industrialization on the growth of large cities in Taiwan were much stronger than the direct effects insofar as industrialization exerts its

TABLE 3.1

THE GROWTH OF MANUFACTURING AND SERVICE EMPLOYMENT

	Annual Growth Rate of Employment			
	Manufacturing		Service	
	Five Largest Cities	Rest of Taiwan	Five Largest Cities	Rest of Taiwan
1950–55	6.9	5.5	4.2	0.7
1955–60	4.0	6.3	3.2	3.6
1960–65	3.2	4.2	4.0	3.1
1965–70	-1.0	2.5	10.3	8.7
1970–75	13.5	20.6	6.5	5.1
1975–80	8.0	14.1	4.4	4.5
1980–85	3.8	4.4	5.4	3.7

influences via the induced growth of service industries rather than through the direct effect of the increase of industrial workers in big urban centers. This relation is shown in Figure 3.1.

In contrast to the spatial relation between industrialization and the growth of urban areas, the temporal influences of industrialization on urbanization for the nation as a whole may be different in strength and direction. This is mainly due to the fact that statistical measurements used for areal units at any particular point in time are different in meaning from the same measurements used at the same time for the whole nation. As industrialization proceeds over time, population tends to be assembled in urban centers to take advantage of technical efficiency and services. Normally the process of industrialization was accompanied by the accelerated growth of tertiary industries, and hence contributed to further urban growth.

In the developing countries today, however, this relation is likely to be obscured by the pressure exerted by rapid growth of the urban population, caused by both natural increase and the receipt of migrants from rural areas. The increased urban population would mostly be absorbed but not necessarily fully or efficiently utilized by the traditional sectors in the tertiary industries. As manufacturing activities expand, needs of the accompanying expansion of tertiary employment are in part met by recruiting of new members of the labor force, but these positions are largely met by more fully and efficiently utilizing workers already employed in the sector. The magnitude and direction of the effects of this sort of structural change of labor on the process of urbanization thus would largely depend on the extent of labor underutilization in the tertiary industries and the spatial distribution of newly recruited members.

In Taiwan, retail stores are open from morning to late evening seven days a week. Employees, persons who are self-employed, and family helpers work or stand-by as long as the store in open. In terms of working hours this amount of performance no doubt represents full labor utilization, but in terms of efficiency there is plenty of room for improvements in productivity, either by increasing the amount of actual work, or by transferring some of these people to more productive jobs.

According to a migration survey in central Taiwan in 1967, 71 percent of the migrants to the central city of the Taichung metropolitan area who were currently employed in tertiary industries had their previous jobs in the same occupation in rural areas; and only 11 percent of them changed from being farmers (Speare, 1969). If these findings are representative, we may infer that any increase in absolute or relative numbers in service employment in the labor intensive stage of development of Taiwan would not necessarily be positively associated with the increasing rate of urbanization.

Technological progress and changes in level and pattern of external and domestic demand are generally considered to be the principal determinants of industrialization. Taiwan, in the 1950s like all other developing countries in the earlier stage of its industrialization, was endowed with abundant unskilled manpower, but suffered from a scarcity of capital. The choice of appropriate technology in this situation inevitably tends to be labor intensive or capital saving.

The strength and direction of the effects of this type of industrial technology on urbanization were different for the agricultural and nonagricultural sectors. In the nonagricultural sectors, introduction of and improvements in labor intensive techniques created more new employment opportunities than would have resulted from capital intensive technology. Therefore there was a direct and positive influence on industrialization and thus on urbanization.

The experience of Western Europe in the nineteenth century and in Taiwan in the recent years indicates that improvements in agricultural techniques resulted in a marked rise in the level of living and hence a substantial increase in agricultural surplus and effective demand for manufacturing goods and services, thus fostering industrial development. In the Western countries the improvements in agricultural techniques were mainly capital intensive, and drove farm workers off the land and constituted a direct stimulus for rural-to-urban migration. In Taiwan, however, the improved agricultural techniques in the last two decades were mainly concentrated on such labor intensive methods as introduction of new high-yield plant varieties. Furthermore, the land reform program favored the adoption of labor intensive techniques, due to its encouragement of small holdings of land. The average farm size was 1.4 hectares in 1950, when the land reform program started, and continuously declined to an average of a little less than one hectare in the 1970s.

All of these factors tended to keep farm workers in rural areas instead of releasing them to urban sectors. This type of technological effect on rural development is likely to occur in the developing countries in the near future if the newly started green revolution is accompanied by land reform programs which encourage small farms in these areas. This experience in Taiwan may be more relevant to what may happen in developing countries than the experiences of the Western countries. Today improvements in agricultural techniques may exert an immediate negative influence on urbanization on the one hand but on the other hand may have an indirect influence on urbanization by its effect on industrialization.

Changes in the level and pattern of external and domestic demand constitutes another major factor in affecting industrialization and urbanization. Since Taiwan is a small island with a linked domestic market, its development has to rely heavily on the international market. Since Taiwan's cultivable land has been fully utilized since the 1960s and agricultural technology has also attained a relatively high level, there is not much room for the expansion of agricultural exports after feeding its rapidly growing population. The only alternative then is the expansion of manufacturing goods. The foreign demand is primarily subject to changes in external conditions, and thus is treated as an exogenous variable in our model.

The level and pattern of the domestic demand for manufacturing goods are, according to Engel's law, directly related to the level of per capita income. The size, distribution, and rate of growth of population can have adverse effects on the growth of per capita income in the developing countries (Coale and Hoover, 1958). An elaborated study for Taiwan has also come to the same conclusion (Liu, 1973). We thus treat the population factors as the principal endogenous variables in the determination of domestic demand, in addition to income.

Education is hypothesized to be the other intermediate variable with a positive and direct influence on urbanization in Taiwan, spatially and temporally. Evidence of educational selectivity of rural-to-urban migration could be interpreted to mean that the higher the educational attainment of the population in an urban place the higher the tendency to provide more educational and economic opportunities to attract persons from other places, and also the better the chance of holding and educating them.

The argument for a temporally positive causal influence of education attainment on urbanization may run as follows. The high Chinese traditional value given to intellectuals and to improving one's education is still highly respected in Taiwan. People thus are highly motivated to pursue better education irrespective of the needs of the economy. Primary education had become universal by the late 1950s and junior high education has been compulsory since 1968. The technique and management needed in the current small family farm can be efficiently and appropriately handled by persons with primary education. Thus, those with more than primary schooling are over-qualified and tend to seek nonfarm employment in urban centers accelerating thereby the rate of urbanization.

The third relevant intermediate variable with an overall but independent influence on urbanization is the type of socioeconomic organization. The traditional extended family and firms organized on the basis of family and clan are still prevalent in both rural and urban Taiwan. The family-centered society is likely to be a hindrance to the process of urbanization and hence to industrial development.

Consequences of the effect of urbanization have been manifested in numerous cultural, socioeconomic, and demographic aspects. The most immediate and significant consequences in our model are demographic. In general, urban areas exert a differential pull on persons of different sexes, ages, and socioeconomic statuses, on the one hand, and have a profound effect on demographic behavior concerning growth and formation of a family, on the other. The sex ratio, age structure index, and average size of households are directly related to urbanization. The fertility level is also associated with urbanization through the effect of urban residence on marriage patterns, on desired number of children, and on contraceptive usage.

The changing demographic variables resulting from urbanization in turn effect the level and pattern of domestic demand. This completes the chain relationship model of urbanization.

It should be noted, however, our model presents only one side of the coin--the urban; and leaves the other side--the rural or agricultural sector--in the background without being explicitly treated. This does not mean that the rural sector should be neglected or is of secondary importance. It is well established that improvement in agricultural techniques and thus production is the engine of industrialization and urbanization in industrially advanced countries and the problems of over urbanization of the present-day developing countries is the result of over-population (United Nations, 1973:201). Accordingly, if we look at the development of the nation as a process of transformation of the social and economic structure from agricultural to industrial or from rural to urban, our model should be the one for the nation as a whole.

FINDINGS

The theoretical model developed in the last section lends itself to multivariable techniques of analysis. Since a series of causal stages among several variables has been implied, the method of path analysis seems appropriate. Each causal stage is assumed to be a sub-recursive model which will not include feedback efforts, therefore conventional least squares procedures can be employed to calculate the size of the paths. Additional assumptions implied in the method of path analysis are that: (1)

multicollinearity is low, (2) effects between variables are linear and additive, and (3) residuals are uncorrelated with any independent variables directly affecting the dependent variable upon which it acts (Heise, 1969).

The measurement of variables specified in the theoretical model are as follows:

Technology	X_1 = time trend, 1952 = 1, 1952 = 2, etc.
External Demand Pattern	X_2 = proportion of industrial products to total exports
Internal Demand Pattern	X_3 = share of nonagricultural consumption in total private consumption
Industrialization	X_4 = proportion of industrial workers to total employment
Tertiary industries	X_5 = proportion of tertiary workers to total employment
Institutional factors	X_6 = proportion of wage earners (employed workers) to total employment
	X_7 = posting of domestic letters per capita
Educational attainment	X_8 = percent of junior high graduates and higher in total population aged 15 years old and older
Urbanization	U_1 = proportion of internal migrants to specific urbanized areas, 1972
	U_2 = ratio of central city rate of growth to the national rate
	U_3 = ratio of metropolitan rate of growth to the national rate

Three measures of urbanization are used. First, the proportion of migrants (excluding migrants from mainland China during early 1950s) in specific urbanized area (U_1) is chosen to measure the level of urbanization of each urban area in the analysis of the causes and the effects of the urban hierarchy in Taiwan.[1] The second and third measures are the ratio of the central city rate of growth to the national rate (U_2) and the ratio of the metropolitan rate of growth to the national rate (U_3). These are used to measure the rate of urbanization of Taiwan as a whole in the longitudinal analysis. These types of measures have been used widely in studies on urbanization. Although they are not ideal because they ignore urban and rural fertility and mortality differentials, they have the advantage that the data are available for a consistent set of metropolitan areas for the period under study.

Cross-sectional Findings

The path diagram of the model showing values of path coefficients derived from the 1972 and 1980 cross-sectional data on 42 urbanized areas is presented as Figure 3.2.

The causal-step of the determination of the rate of industrialization was eliminated from the model because the cross-sectional analysis is appropriate for dealing with location of industries, but not with the process of industrialization.

The level of urban growth of the cities, as measured by the proportion of migrants in the urbanized area, depends directly on the levels of tertiary workers in employment, of employed persons, and of educational attainment. The influence of industrial development, represented by the proportion of industrial workers in total employment, on urban concentration comes mostly through its effect on tertiary employment. This is seen from the fact that in the phase of labor-intensive development around 1972, the largest path to the variable of urbanization is from the proportion of tertiary employment (0.37). The direct path from proportion of industrial employment to urbanization is negligibly small (0.07); This evidence seems to support our hypotheses about the spatial relation of growth of urban areas and distribution of economic activities. Under the severe population pressure on land, tertiary industries tend to concentrate in large metropolitan areas while manufacturing tends to spread out in small urban places. As industrialization and urbanization proceed toward higher levels, this relation tends to strengthen. By 1980, the direct path from proportion of industrial employment to urbanization is negative. At the same time, the influence of tertiary employment on urbanization becomes much stronger.

The second largest path to the urbanization variable in 1972 is from

----

[1] In general, the population size of urban places has been a popular and commonly-used measure. We do not use this measure in our analysis because studies on city systems tend to lead to the conclusion that "there is no relationship between type of city size distribution and the degree to which a country is urbanized" (Berry, 1961, p. 587).

FIGURE 3.2

PATH DIAGRAM FOR URBANIZED AREAS IN 1972 AND 1980

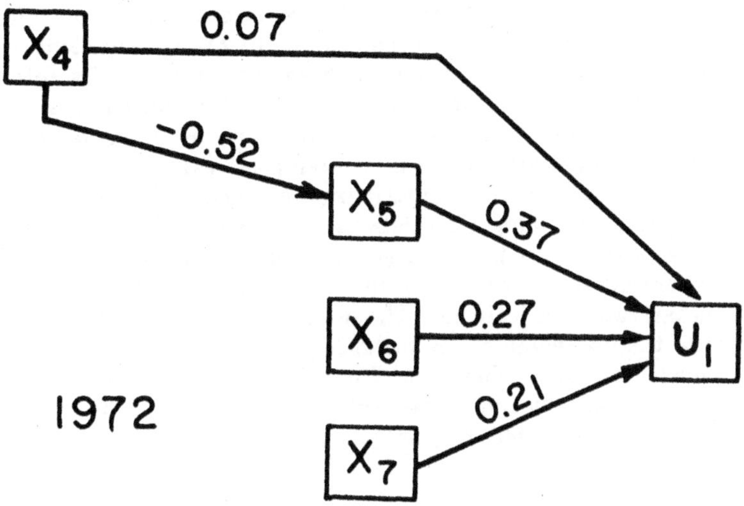

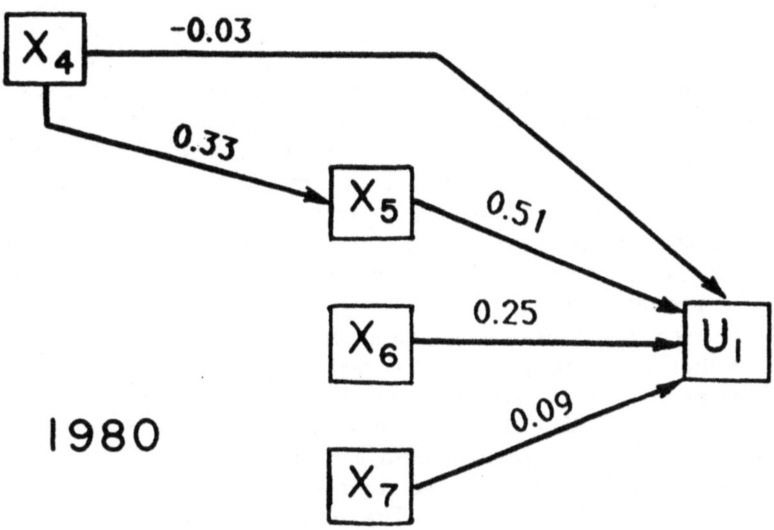

the proportion of wage workers in total employment (0.27). If this variable can appropriately measure the degree of modernization with respect both to types of economic organization and attitude of employers toward hiring nonfamily related persons, the relatively positive path again supports the notion that the family-centered economy is a hindrance to the speed of urbanization, and that nonfamilial employment accelerates urbanization.

Although the direction of the path coefficient from educational attainment of urbanized areas to urbanization (0.21) is consistent with hypothesis of a positive relation between these two variables, its relative importance is not as large as had been expected. Furthermore, in 1980 this relationship becomes weaker (.09). As will be seen later, the magnitude of educational attainment in the temporal causal relation ranks as the second largest, next to industrialization. The different effects of education in spatial and temporal relations may indicate that the specific flow of better-educated persons into large cities was not as important as the universal expansion of education in the determination of recent rapid growth of urbanization in Taiwan. The weak relation observed in 1980 may be due to the upgrading of the technology and government policies to disperse industrialization. The four immediate explanatory variables together predict 35 percent and 57 percent of the variance in the degree of urbanization in 1972 and 1980, respectively.

Longitudinal Findings

In this section, social, economic and demographic data for Taiwan for the period of 1952 to 1973 were used to approximate the causal relationship model advanced in Figure 3.1. Values of the path coefficients obtained from the use of the ratio of the growth rate of central cities to the national rate (U_2) and the ratio of metropolitan growth to the national rate (U_3) are shown respectively in Figure 3.3 and Figure 3.4. It is seen that paths to the index of urbanization for metropolitan areas are generally stronger than those measured in terms of central cities while the directions of the relationships are identical. This implies that the process of urbanization in Taiwan has tended to disperse into a large territory surrounding the largest cities, rather than to concentrate in the central cities; otherwise the cause and effect relations are the same for central cities and entire metropolitan areas.

One of the advantages of longitudinal analysis over the cross-sectional analysis is that the former permits us to test the part of the model relating to the determination of the rate of industrialization. The results of the path analysis here indicate that changes in internal demand for manufactured products have the largest positive path contribution (0.97) to the recent industrialization in Taiwan while the contribution of changes in external demand (0.06) is negligibly small. The probable interpretation of this significant difference between these coefficients is that the expansion of industrialization in Taiwan had relied heavily on the domestic market by

FIGURE 3.3

PATH DIAGRAM FOR URBANIZATION (CENTRAL CITIES IN TAIWAN) 1952-1973

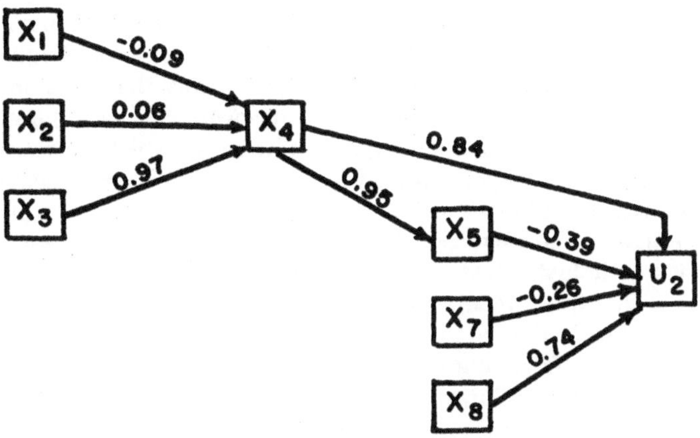

FIGURE 3.4

PATH DIAGRAM FOR URBANIZATION (METROPOLITAN AREAS IN TAIWAN) 1952-1973

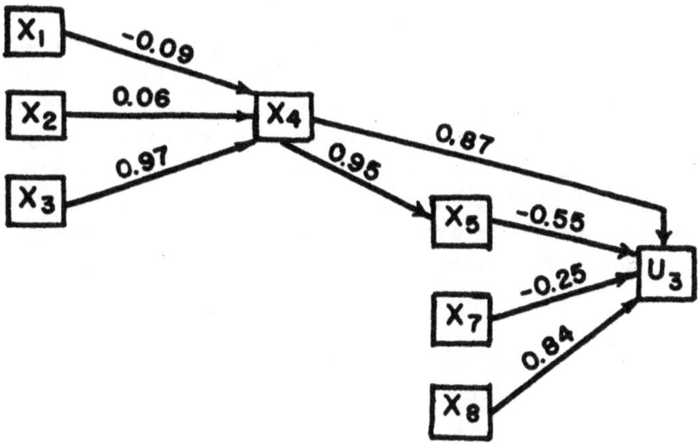

the use of import substitution policy throughout the whole observed period, while the reliance on export expansion began late in the 1960s and is important for only a short portion of the analysis period. The rate of industrialization in this study is measured in terms of industrial employment, which introduces a bias against the contribution of export expansion in comparison with the other measures, in terms of value added. When the import substitution industries started to grow, the abundance of labor, in contrast to the shortage of capital and entrepreneurship, led toward the mushrooming growth of small capital-saving family firms, and thus absorbed a large share of the surplus labor force. The late entry of the exporting industries confronted them with a relative scarcity of labor--especially the particular kind of skilled labor needed, but they had more accessible sources of capital and technical knowledge. The changing situation channeled the expansion of export industries, which were still labor-intensive, towards large, relatively capital-intensive modern incorporated enterprises (Hsing, 1971). Its capacity for labor absorption was inevitably limited. The negative, but weak, path coefficient from the technological index to industrialization (-0.09) may be also an indication of these patterns. The three independent variables together explain 87 percent of total variance in the rate of industrialization.

The next causal step following the determination of industrialization is the examination of the constellation of variables determining the rate of urbanization. The respective path coefficients calculated from the statistical data for Taiwan for 1952-1973 are all pronounced and statistically significant. Basically, the strength and direction of path coefficients from the proportion of industrial employment and proportion of tertiary employment directly to urbanization, and the indirect effect of industrial employment via its effect on tertiary employment, are not the same as the spatial path coefficients. While the cross-sectional analysis showed larger metropolitan areas to have higher proportions in the tertiary sector than smaller areas, the time series analysis shows a strong direct effect for industrialization. This suggests that industrialization is the main determinant of how fast the national level of urbanization increases in a particular time period. Industrialization is highly correlated with the size of the tertiary sector, as indicated by the coefficient of .95 between X_4 and X_5. This high degree of multicollinearity makes it hard to separate the effects of these two variables. However the negative coefficient between X_5 and urbanization suggests that there is a negative effect on urbanization at the national level when tertiary growth exceeds that expected on the basis of industrialization.

Due to the lack of time series data on wage earners, in the longitudinal analysis this variable was replaced by the posting of domestic letters per capita. The direct path coefficient between letters per capita and urbanization (-0.26) is negative and significant. This may support the idea that people who are able to utilize correspondence through the mails more frequently will be better informed about conditions in urban labor markets, and could move to cities more efficiently, hence avoiding the unnecessary expansion of urban areas. Educational attainment which turns out to have the second largest direct positive path coefficient to urbanization (in contrast to being the smallest in the cross-sectional analysis) seems to imply that the relative educational level of particular urban centers is not as important as the rise in educational level of the entire population in fostering recent rapid urban growth.

Out of the total variance in the rate of urbanization, 91 percent is explained by the four variables. Yet, comparing this with the 35 percent to 57 percent explained variance in cross-sectional analysis should not lead one to conclude that there is a great improvement in the predictive power of the longitudinal model. The explained variance is unduely inflated by the lack of substantial variation in variables over a relatively short time period.

CONCLUSIONS

Both our cross-sectional and longitudinal analyses for Taiwan show the existence of a strong relationship between development and urbanization. In spite of the severe population pressure on land and shortage of capital at the initial stage of development, Taiwan's economy has experienced a "virtuous" circle of cumulative growth. Briefly, the mechanism is as follows: As demand patterns changed in favor of industrial goods and as the needed technology is available, the growth of modern industrial activities is stimulated and gains momentum. The induced industrialization, at first, brings in an accompanying growth of tertiary industries over time (spatially it causes growth of the modern tertiary industries but a depression of the traditional ones in large cities). Together with necessary changes in institutional conditions and an increase in the educational attainment of the general population, industrialization and the accompanying expansion of tertiary sectors foster a process of population concentration in urban areas--urbanization.

Partly via its inverse effect on proportion of married women, but mostly through its influences on the transformation to new values and motivations, urbanization facilitates a decline in the birth rate and hence decreases the rate of population growth and the economic burden of excess children. Furthermore, the reduction of population pressure increases per capita income and hence, through the working of Engel's law, increases the demand for more industrial goods. This completes the causal circulation of urbanization. In view of the strength and direction of the path coefficients connecting the causes and effects obtained from the statistical data of Taiwan, it may suggest that an initial surge of any of the variables in the causal interrelationship would set off a series of sequential reactions which may reinforce or counteract it and give rise to a cumulative upward or downward process. It should again be stated that although our model deals directly with urbanization, the working of the rural aspects is intimately associated in the background, and hence this model of uni-directional relationships could represent the development of the nation as a whole.

The significance of the findings must be tempered by the recognition that the statistical analysis has been based on empirical data for Taiwan for a small number of spatial observations and over a relatively short time period. The model presented in the present chapter represents only one theoretical interpretation of the causal relationships between urbanization and socioeconomic development. The model could probably be improved by adding additional measures of development. Yet, the analyses point clearly to a strong association between development and urbanization both among urban areas at particular points in time and for the entire country over time.

4

The Growth of the Taipei Metropolitan Area

The previous chapters have shown that population increase in the major cities of Taiwan slowed during the 1970s, while the intermediate and small cities continued to grow. It was also found that the shift was most noticeable in the Northern Region which contains the capital city of Taipei. Utilizing registration and survey data, this chapter examines the patterns of population change in the various parts of the Taipei Metropolitan Area. Taipei is as an example of urban expansion and suburbanization in the developing world.

Taipei, the island's largest city, is located in an alluvial basin in northern Taiwan. Surrounded by hilly land, Taipei itself is only 150 to 600 meters above sea level. The triangular basin enclosing the city is about twenty-five kilometers from north to south and encompasses about 400 square kilometers. It is drained by the Tamshui River and its tributaries -- the Keelung River from the northeast, the Hsintien River from the south and the Takokan River from the southwest. The Tamshui River flows out of the basin at the northwest corner, and empties into the Taiwan Straits.

HISTORICAL PERSPECTIVE

Prior to 1600, the Taipei basin was sparsely occupied by the tribe of Kelagalan. Only occasionally did fishermen and merchants from mainland China, using the mouth of the Tamshui River as rest and supply base, visit the basin. In 1626, the Spaniards came to northern Taiwan and founded a trading post in Keelung harbor. Three years later they founded another post at the mouth of the Tamshui River. The Dutch, with established strongholds in southern Taiwan, took over the Spanish trading posts in the north in 1642, and remained there until 1667, six years after their compatriots in the south surrendered to the Chinese general Cheng Ch'eng-kung. Neither the Spaniards nor the Dutch extended their conquest into the hinterland; hence, no significant urban or rural settlements developed under their rules.

In Cheng's period (1661-1683) part of the lower course of the Tamshui River was gradually cultivated and settled by his followers who had come directly from south China. Some fifty years after the Ch'ing Dynasty established a government succeeding Cheng's, the Taipei basin was virtually fully occupied. The occupants were partly Chinese migrants who had moved directly from the mainland China province of Fukien, and partly Chinese settlers from southern Taiwan. A market town of Meng-chia, the original core of Taipei city, was established along the eastern bank of the Tamshui River between the two tributaries, the Hsintien River and the Keelung River. This town served as a trading center for the new agricultural area, linking it to the nearby aboriginal area and mainland China. By 1853, with a population of about 18,000, it ranked as the third largest city of Taiwan. As a result of the 1851-1861 conflicts between two groups of settlers, the defeated group moved a short distance downstream to the north and built a new market town known as Ta-tao-chen. This town soon became the domestic and international trade center of Taiwan, because the silting of the Tamshui River had made Meng-chia inaccessible to seagoing junks.

In 1875, in recognition of the importance of Taiwan and the strategic position of Taipei, the Ch'ing government established Taiwan as a prefecture of Fukien province and built a walled inner city as the seat of administration. In 1885, Taiwan became one of the eighteen provinces of China and the inner city became the provincial capital. Due to the rapid silting of the Tamshui River within a short period the whole course became unnavigable for big junks. Keelung, a deep gulf with a rocky shore 26 kilometers away from the inner city of Taipei, was selected as the trade outlet for the Taipei basin. A railway was laid at that time to connect the inner city of Taipei with Keelung in the north and Hsinchu in the south. This first railway made the Taipei basin easily accessible to the sea, as well as to the vast fertile plain of western Taiwan.

When Taiwan was ceded to Japan in 1895, Taipei was chosen as the administrative center. The city wall was torn down and the boundary expanded to include the two old market towns which had been outside the walled city. In 1895, the population within these new boundaries was 47,000. Since then, the city has become the largest political, economic and cultural center of Taiwan.

Taiwan was returned to China in 1945 and Taipei remained its provincial capital. In 1950 the central government evacuated from mainland China, and Taipei became the temporary capital of the Republic of China. The government of Taiwan province moved to central Taiwan in 1956. In 1967, Taipei was made a special city under direct jurisdiction of the central government. It is now one of the major cities in east and southeast Asia.

Since Taipei is bounded by rivers on the north, south and west, later expansion of the city first took place southward and northward along the Tamshui River, then moved eastward after reaching the boundaries of the

FIGURE 4.1

TAIPEI METROPOLITAN AREA AND COMPONENTS

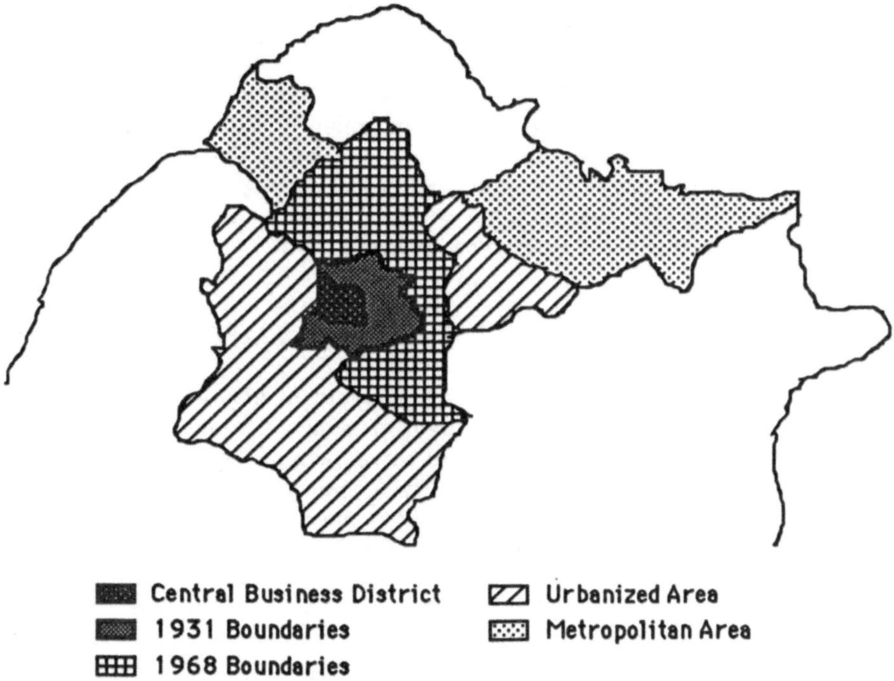

other two rivers. The four old precincts of Taipei, which are the central business district today, comprised an area of 18.65 square kilometers. Extended to cover the adjacent suburbs in 1932, the city then contained a total area of 66.99 square kilometers. The boundaries remained unchanged until 1968, when they were again extended to include the Yangmin Shan administrative area which contained some of the national government, and four additional townships in the east. Since then, the total area of the city has been 272.14 square kilometers. In the north and east, the new boundaries extend to the ridges of the basin; in the west and south, the boundaries are still the nonnavigable rivers. Due to rapid population growth, industrial expansion and particularly the construction of large bridges across the rivers, the plains to the west and south of the city have been connected and physically integrated with Taipei city.

In order to discuss urban development and regional planning, it is necessary to also consider the neighboring urbanized area outside the city. According to the definitions adopted in 1975 (Liu, 1975), the Taipei urbanized area was defined to include the city of Taipei and 12 surrounding cities and townships which were highly urbanized. The metropolitan area included the entire urbanized area plus the ports of Keelung and Tanshui and the town of Jui-fang. The total urbanized area contains 672 square kilometers.

The Taipei Metropolitan Area is an agglomeration of several administrative units: (1) Taipei Municipality as the central city, (2) the provincial city of Keelung, and (3) several smaller cities and townships of Taipei County which surround Taipei Municipality (Figure 4.1). The total metropolitan area has an area of 945 square kilometers. Within Taipei Municipality itself, three rings can be conveniently distinguished according to the history of changes in the city boundaries. The central business district, which comprises four precincts, was the original core of the city. The second ring includes territory which was added to the city in 1931, while the third ring includes territory added in 1968.

RAPID POPULATION CONCENTRATION IN TAIPEI

Because Taipei was the center of government of Taiwan both during the Japanese occupation and since then, movements from Japan and other parts of China have been significant in the growth of the city. According to the register, at the end of 1905 the population residing within the 1913-1958 city boundary was 107,616. It grew steadily at a moderately fast rate of about 4 percent per year until 1940, when the population reached 353,744. After World War II there was a sudden sharp drop due to the repatriation of 100,000 Japanese, but by 1950 the loss was more than compensated for by an influx of Chinese from mainland China. Since 1950, the population of Taipei has continued to grow at a high rate of more than 5 percent per year; only in recent years has the growth rate gradually approached a moderately low level of 3 percent per year.

The ethnic origins of the population can be approximated by using the concept of domicile in the household register. Domicile refers to the place of origin of one's ancestors and children usually take the domicile of their father. In Table 4.1 the domiciles of Taipei residents within the 1931-1968 city boundaries are shown. Prior to the Second World War the majority of Taipei residents were Taiwanese, people who had moved to Taiwan from mainland China before 1895 and their descendents. The proportion of Taiwanese who had domiciles elsewhere in Taiwan and had moved into Taipei city after 1895 remained stable at an insignificant 6 percent throughout the period 1920-1940. However, their share increased rapidly from 14 percent in 1950 to 40 percent in 1980. This growth clearly indicates that rural to urban migration in Taiwan became a significant phenomenon only since World War II.

During the period of Japanese occupation (1895-1945), the migratory flow from mainland China to Taiwan was strictly limited to 10,000 seasonal workers per year. In this half-century, mainland Chinese never

TABLE 4.1

ORIGINS OF TAIPEI RESIDENTS 1920-1980
(Using Fixed 1931 Boundaries)

	Percentage Distribution					
	Taiwanese					
Year	Taipei Origin	From Rest of Taiwan	Mainland Chinese	Japanese and other Aliens*	Total Percent	Total Number ('000s)
1920	63.4	5.2	5.4	26.0	100.0	177
1930	60.1	6.2	6.1	27.6	100.0	255
1940	61.7	6.0	3.8	28.5	100.0	354
1950	52.7	14.1	33.1	0.1	100.0	503
1960	38.9	23.0	37.8	0.2	100.0	901
1970	33.5	27.9	38.0	0.6	100.0	1383
1980	28.7	40.4	30.9	NA	100.0	1474

*Other Aliens accounted for 0.1% of the population up to 1960. Thereafter they have grown to 0.4 percent.

exceeded 6.2 percent of the population of Taipei. They were concentrated in two occupation groups, merchants and laborers in small manufacturing or transportation.

In spite of vigorous efforts of the Japanese government to persuade Japanese farmers and fishermen to settle in Taiwan, more than 95 percent of the active male immigrants from Japan were engaged in nonagricultural activities. Fully half of them were government officials and professionals. Among all Japanese immigrants about one third settled in Taipei city. Even within the city the settlement was concentrated, with a core of only Japanese in the inner city, surrounded by a substantial area where the Japanese predominated. Practically all Japanese were evacuated from Taiwan in 1945. By 1950, the vacancies had been overfilled by a sudden inflow of mainland Chinese. Thereafter the continuous inflow of descendents of pre-1895 migrants, mainly from rural areas of Taiwan, made them an increasingly important component of the population of Taipei. By 1980, their share was slightly higher than that of the other two major population groups, i.e., descendents of pre-1895 Taipei residents, and mainland Chinese migrants and their descendents.

Unlike some western cities, Taipei and the other cities and towns in Taiwan were not structurally formed around a purely commercial central business district. Instead, a primary commercial-cum-residential area which performs the functions of the western central business district developed in each city or town. Juxtaposition of shop and residence was primarily due to the prevalence of the traditional type of business organization. In Taiwan, the traditional business firm is small in scale and largely owned, managed and operated by members of a single family. For purposes of full participation, the most economical and efficient arrangement is to use the ground floor of the building for the commercial or service function and use the rear and/or upper levels for homes of the owners and their families.

Throughout most of its history the central commercial area of Taipei city has been densely populated with permanent residents. This situation, however, has been gradually changing in response to the modernization of business organization, improvement in the use of public and private transportation and the rising standard of living. As economic prosperity of the central business district continued, the size of its registered resident population increased only moderately from 1920 to 1970, and thereafter decreased (Table 4.2).

From the 1960s, there were many projects to renew substandard dwellings in the central business districts which included some of the oldest housing, accompanied by various programs of developing new communities in the outer rings of the city. These are some possible explanations for the sudden and temporary, but striking rate of population decrease observed for 1970-80. As the center of Taipei City moved toward the newly expanded areas on the east part, the original CBD retained some of its importance by being filled with retail businesses.

Table 4.3 further indicates that the net out-migration from the district was 3.6 times the amount of its natural increase in 1973-78. The decline of this ratio to 2.6 times for 1978-83 indicates the diminishing relative importance of net migration in population change. This finding is also evidenced by the substantial decrease in the number of net out-migrants from the CBD (from 57,365 to 31,724).

The areas added to Taipei City in 1931 grew faster in terms of population than the current municipal areas during the years before 1970. In particular, the average annual growth rate of 8.3 percent for 1960-70 is the highest found in Table 4.2. After 1970, however, the increase rate became lower than that of Taipei Municipality and has stayed at 1.5 percent since then. Table 4.3 further shows that net migration offset one third of the size of natural increase observed for the areas in the years after 1973. This is probably due to the full development of the pre-1931 areas and the new expansion to the areas added to the municipality in 1968.

The six precincts added to the city in 1931 are important parts of the municipality. As the original CBD became old and crowded, the center of the city started to move eastward in the 1960s. While the six precincts still have a lot of residential communities, two of them (Chung Shan and Sung Shan) have become new centers of commerce, finance, and modern

TABLE 4.2

GROWTH OF THE TAIPEI METROPOLITAN AREA, 1920 TO 1985

Population (in 1000s)

Year	Central Business District	Areas Added in 1931	Areas Added in 1968	Total Taipei City	Rest of Urban Area	Rest of Metro Area	Total Metro-politan Area
1920	109	66	49	224	116	89	428
1930	131	115	56	302	128	126	555
1940	*	326	66	392	147	188	726
1950	213	291	112	616	210	221	1046
1960	255	644	198	1097	429	333	1860
1970	262	1113	394	1770	853	447	3070
1980	179	1295	746	2220	1854	477	4551
1985	173	1397	938	2508	2227	484	5219
	Percentage Annual Growth in Preceding Period						
1930	1.8	5.7	1.4	3.0	1.0	3.6	2.6
1940	*	2.9	1.6	2.7	1.4	4.1	2.7
1950	*	4.4	5.5	4.6	3.7	1.6	3.7
1960	1.8	8.3	5.9	5.9	7.4	4.2	5.9
1970	0.3	5.6	7.1	4.9	7.1	3.0	5.1
1970	-3.7	1.5	6.6'	2.3	8.1	0.6	4.0
1985	-0.8	1.5	4.7	2.5	3.7	0.3	2.8

*Combined with population in areas added in 1931.

**Based on the formula: r = 100*(EXP(ln(P2/P1)/n)-1), where n = no. of years.

Sources: 1940 to 1970 from Paul K.C. Liu, (1979). 1980: 1980 Taiwan-Fukien Demographic Fact Book. 1985: 1985 Taiwan-Fukien Demographic Fact Book.

TABLE 4.3

DECOMPOSITION OF POPULATION CHANGE FOR TAIPEI METROPOLITAN AREA, 1973-1983

	Population Change	Natural Increase	Net Migration
1973-1978			
1. C.B.D. (Areas before 1931)	-41,218	16,147 (-39.2)	-57,365 (139.2)
2. Areas added in 1931	67,255	101,817 (151.4)	-34,562 (-51.5)
3. Areas added in 1968	179,172	58,242 (32.5)	120,930 (67.5)
4. Current Municipality (=1+2+3)	205,209	176,206 (85.9)	29,003 (14.1)
5. Surrounding Urbanized Areas	555,264	153,975 (27.7)	401,289 (72.3)
6. Surrounding Metropolitan Areas	11,124	38,466 (345.8)	-27,342 (-245.8)
7. Total Metropolitan Area (=4+5+6)	771,597	368,647 (47.8)	402,950 (52.2)
1978-1983			
1. C.B.D. (Areas before 1931)	-19,721	12,003 (-60.1)	-31,724 (160.1)
2. Areas added in 1931	59,093	91,995 (155.7)	-32,902 (-55.7)
3. Areas added in 1968	185,397	70,262 (37.9)	115,135 (62.1)
4. Current Municipality (=1+2+3)	224,769	174,260 (77.5)	50,509 (22.5)
5. Surrounding Urbanized Areas	491,457	206,113 (41.9)	285,344 (58.1)
6. Surrounding Metropolitan Areas	11,542	36,717 (318.1)	-25,175 (-218.1)
7. Total Metropolitan Area (=4+5+6)	727,768	417,090 (57.3)	310,678 (42.7)

Note: Figures in parentheses denote the relative weights in percentages of natural increase and net migration to population change.

service industries. Another two precincts (Ta-an and Ku-ting) are characterized by mixtures of residential, commercial, and educational areas.

The areas annexed to Taipei Municipality in 1968 did not grow rapidly until after 1940 (Table 4.2). In the decade of the 1960s during which the annexation occurred, their growth rate reached the highest level of 7.1 percent per annum. Growth remained at a high level of 6.6 percent in the 1970s and then declined to 4.7 percent for 1980-85. Since 1970, this has been the most rapidly growing part of Taipei municipality. According to Table 4.3, net migration accounted for three-fifths to two-thirds of the population growth in the period of 1973-83. The migrants include those from the original CBD, the 1931 added areas, and places outside the municipality. The data along with those discussed earlier clearly indicate a dispersion of population within the municipality as well as an intrametropolitan redistribution.

It should be noted that the six new precincts of 1968 are all large in terms of land area, including substantial amounts of hilly places on the north, east, and southeast sides of Taipei basin. These precincts were developed mainly as residential and recreation areas as well as locations for schools. Given the high potential of these areas for future development, the population growth rate is anticipated to maintain at a higher level than those of other parts of the city.

In the early years, the surrounding urbanized areas were obviously rural in character. The population of these areas did not start to increase significantly until the end of World War II (Table 4.2). During that era, they were areas of out-migration to Taipei. In the three decades following 1950, however, the urbanized areas had remarkably high rates of growth, averaging 7.1 percent to 8.1 percent per year. For 1980-85, the rate declined to 3.7 percent, which is still higher than that of either the entire metropolitan area or the municipality. In the two decades before 1970, the population residing in the surrounding urbanized areas accounted for less than a quarter that of the metropolitan areas as a whole, but by 1985 it had increased to include 43 percent of the population of the total metropolitan area.

The growth of the surrounding urbanized area occurred during the period in which Taiwan became a newly industrializing country. The urbanized areas surrounding Taipei were the locations of many new manufacturing establishments. In the early stage, the secondary industries provided employment opportunities for those who had been pushed out of the agricultural sector due to high population-land ratios resulting from the demographic transition (declining death rates and high birth rates). In latter years, the factories absorbed the excess labor released by agricultural mechanization and by the shrinkage of the primary sector. The surrounding urbanized areas were destinations for many migrants looking for off-farm jobs. As shown in Table 4.3, net migration accounted for 72 percent of the population increase in the urbanized areas between 1973 and 1978, and for 58 percent in the period from 1978 to 1983.

As noted earlier, the surrounding metropolitan areas, those beyond the urbanized area, include Keelung, Tanshui, and Juifang. The first two are harbors with long histories and the last one was a major site of mining in the early years. Before 1940, these three places grew faster than both the total metropolitan population and the population within the municipality boundaries (Table 4.2), suggesting some net inflow of people. In recent years, however, the areas increased their total population at an average rate of less than 1 percent per year. This low rate of growth indicates a substantial amount of net out-migration, which is confirmed by the estimated net migration for 1973-78 in Table 4.3.

These three surrounding metropolitan areas are all places which developed rather early, and at least two of them have quite specific limitations to growth. Before the 1970s, the population of Keelung City seemed to have reached a maximum amount given geographic constraints imposed by the steep hills which surround it and easy access from Taipei Municipality. Juifang has been stagnant in terms of development since the closing of mines in the 1960s.

The age-sex selectivity of the movers to and from the different areas of Taipei is demonstrated by the distribution of age-specific rates of the estimated net migration for both sexes in Figure 4.2. It is evident that the central business district experienced the heaviest loss of both male and female population of nearly all ages, except for children aged 5-9 and young men 25-29. The greatest losses, however, occurred in the population aged 30 and over. The surrounding metropolitan areas experienced only a small loss in population in the younger and older age groups, while gaining population in the economically productive age groups. The other areas of the Taipei metropolis, that is the 1931 and 1968 affiliated areas and the surrounding urbanized areas, gained population of all ages; the largest gains occurred in age groups 20-44 for men, and age groups 20-29 for women. This large influx of productive-age population into the middle rings of the Taipei metropolis has surely been one of the fundamental causes of the social and economic problems confronting Taipei's government today.

The high rate of natural increase has been an important contributor to the rapid growth of population in Taipei after the Second World War. Due to improvements in public health and medical care, the crude death rate declined continuously from an already low level of 16 deaths per thousand in the late 1930's to 3.6 per thousand in 1975. The birth rate, however, sharply increased after the war because of the baby boom. This high level of births lasted until the late 1950's, after which it declined continuously to a moderate level of about 20 births per thousand population in 1975. As a consequence of these two trends the rate of natural increase of Taipei's population reached a high of 34 per thousand population during 1950-1954, then gradually decreased to a moderately high level of 17 per thousand in 1975.

In contrast to the experience of 19th century western nations, research findings from various less developed countries indicate that natural increase was the chief generator of city growth. For example, Arriaga (1968)

FIGURE 4.2

ESTIMATED NET MIGRATION RATES FOR
TAIPEI BY AGE AND SEX, 1968-72

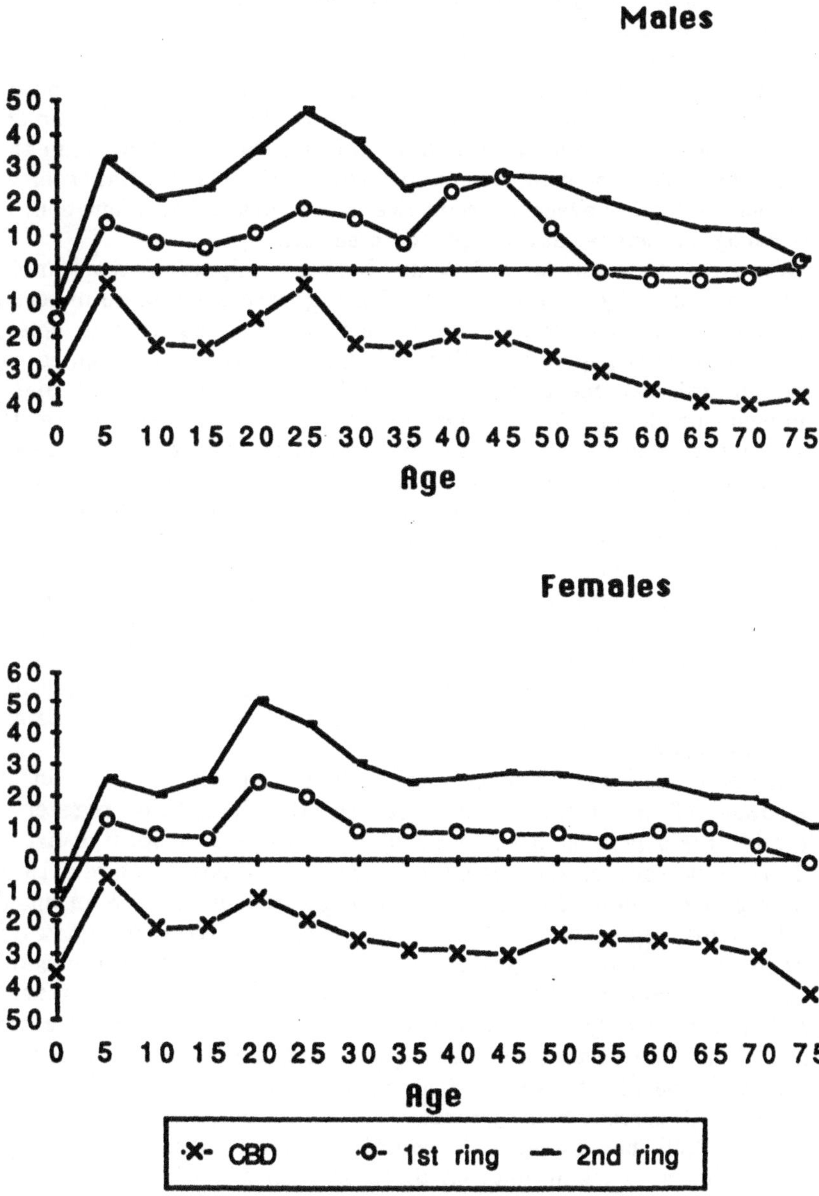

estimates that during the 1950-60 decade, 58 percent of the growth of cities (20,000 or more inhabitants in 1960) in Mexico, 66 percent in Venezuela and 70 percent in Chile was due to natural increase. In the case of Thailand, Goldstein (1971, 1972) reports that in the earlier years of 1947-1967, migration provided the major source of Bangkok's growth, but in the later years natural increase accounted for more of the growth than migration. In the Philippines, although migration had a significant impact (over 50 percent) on the growth of the bigger cities in the 1960s, natural increase was more important than migration to the growth of the smaller cities (Pernia, 1977). All these findings support Davis' (1965) claim that, contrary to popular opinion, the main factor in the rapid inflation of city population in newly developing countries is not rural-urban migration, but sheer biological increase at an unprecedented rate.

However, the simple dichotomy between migration and natural increase in discussing the components of city growth neglects an important segment of the total migration contribution, namely, the natural increase among the migrant population. City natural increase is the weighted sum of the natural increase of in-migrants (after their arrival) and that of lifetime city residents. To the extent that this point is ignored, the relative contribution of migration is underestimated and that of natural increase is overestimated (Weller et al., 1971). In the rest of this section, city natural increase will first be considered as one component and then broken down into two parts: (1) the natural increase of city natives, and (2) the natural increase of in-migrants.

Tsay (1982) computed the components of Taipei's growth between 1968 and 1973 taking account of the possible effects of delayed registration of migrants and the contribution of migrants to the natural increase. A simple division of the population growth showed that net internal migration accounted for 47 percent and natural increase accounted for 53 percent during this period.

However, it was estimated that delays and other errors in the registration of in-migration have caused the household register to undercount the population actually living in the Taipei Metropolitan Area in 1973 by 8 percent of the registered population (Speare, et al., 1975). Assuming that in both 1968 and 1973 the population of Taipei Municipality had the same degree of under-coverage owing to migration registration problems, the actual city growth is obtained by inflating the recorded population growth (342,319) by 8 percent to get 369,705. The difference of 27,386, is then totally attributed to net migration, which reached a total volume of 189,020. Based on these adjustments, the contribution from the net inflow of people is estimated at 51 percent, while natural increase accounts for 49 percent of the growth.

Tsay (1982) then assessed the contribution of in-migrants to the city's natural increase. Lacking data on the proportion of all births in the city that occur to migrant parents and the proportion of all deaths that occur to in-migrants, he assumed that (1) the ratio of in-migration to the city population is constant from year to year, (2) the proportion of married women among women of childbearing ages is the same between in-migrants and city natives, and (3) mortality could be ignored because it is very low.

The proportion of migrants in Taipei's population was obtained from 1973 Taiwan Migration Survey (see Appendix A) This survey showed that migrants accounted for 64 percent of women aged 20-39 in Taipei, while the proportion of natives is only 36 percent. Tsay also used estimates compiled by Liu (1976b) of the mean number of children ever born and the mean number of living children of recent (5-year) migrants with those of long-term migrants and Taipei natives.

These computations indicate that the effective contribution made by in-migrants to the actual population growth of Taipei between 1968 and 1973 is 82 percent (51.1 percent due to net migration per se and 30.7 percent due to natural increase among in-migrants). The natural increase of lifetime city residents accounts for only 18 percent of the growth. As a result, migration seems to have played a more important role than natural increase in the process of the growth of Taipei during the five years from mid-1968 to mid-1973.

The rapid rate of natural increase plus the explosive rate of migration to the middle rings of Taipei metropolis have not only aggravated the urgent demand for urban services, but have also created and intensified problems of social and economic adjustment to a new way of urban life. Since we have described and identified the causes and nature of rapid urbanization of Taipei, we can turn in the following sections to the physical and economic problems that have followed on its heels.

PHYSICAL PROBLEMS

The immediate and urgent problem which confronts the city government of Taipei owing to recent population concentration is how to provide adequate urban services to the population. The target of the first urban plan for Taipei city promulgated in 1903 was to build a modern city to accommodate a population of 150,000. The last revision of the plan during the Japanese occupation was made in 1937 when the size of the target population was increased to 600,000. By 1953, the city's population had exceeded the 1937 planned target, and by 1968 it had doubled its 1953 size. This unprecedented rapid growth forced the urban planning target up to 3,000,000 in 1976 (Urban Planning Department, 1976).

Taipei was seriously damaged by bombing during the Second World War. This physical damage and the postwar influx of migrants from mainland China and rural Taiwan produced a grave shortage of public facilities and housing in the early 1950's. The per capita income then was low, about US $100-150, and the city government resources were stringently limited. It is hardly surprising, therefore, that poor urban physical services, shanty houses and squatters became major concomitants of Taipei's rapid urbanization. Unlike cities in most developing countries, squatter settlements in Taipei are largely located on public land planned for construction of parks, schools, roads and buildings for urban services. The eradication of these illegal settlements in recent years has caused not only great financial strains on the government, but also social disorders and problems. Housing programs sponsored by the city government have been devoted primarily to the relief and the resettlement of those who lost their homes because of squatter and slum clearance. Of the 27,674 housing units built during 1950-1975, 11,425 units, or about 42 percent, fell in this category (Dept. of Public Housing, 1975). The units were small, ranging in size from 25 to 50 square meters, but they are modern and of fairly good quality. The problem, however, is not quality but quantity, because the need significantly exceeded what the government could afford to provide.

As household real income has risen, from NT $70,687 in 1963 (equivalent to US $1,767) to NT $462,854 in 1983 (equivalent to US $11,000), housing conditions have gradually improved. The 1980 Census of Population and Housing showed that the average number of occupants per dwelling unit was 5.2 persons as compared to 5.6 persons for Taiwan as a whole. The floor area per capita was 18.0 square meters in contrast to 15.3 square meters for all Taiwan. Eighty-eight percent of the houses were used for residence only, the rest were used for residence and business combined. More than eighty percent of the housing was built within the last 20 years and about 80 percent had private kitchens, bathrooms and toilets (Population Census Office, 1982). It may be concluded that Taipei's housing standards are relatively good compared with the rest of Taiwan. The ownership of housing in Taipei, however, is fairly low: in Taipei 67 percent of housing is owner occupied, in contrast to 79 percent for all Taiwan. Since the demand for improved housing and home ownership have high priority among the needs of those population groups which have stable occupations and have experienced gradual increases in income, a larger share of total investment of the government's housing programs should be allocated for these lower income groups.

Public utilities have not only kept pace with the rapid population growth but have substantially improved. In 1983, ninety-nine percent of the population was served with electric power, ninety-nine percent with piped water and one-third had a telephone (Table 4.4). The per capita direct consumption of electric power and piped water increased respectively from 316 kilowatts and 67 liters in 1950 to 1,630 kilowatts and 387 liters in 1983. The sharp increase in consumption of electricity, which is supplied by the state-owned Taiwan Power Company, has diverted a large share of the limited energy resources from direct production. Meeting the ever increasing demand for water has also been a serious problem. Available water resources are scarce and have mostly been brought into use. Underground water, which was once a main source of water supply, has to be left idle to prevent the land from sinking. A reservoir in the vicinity of Taipei, with a capacity to serve an additional population of 1.5 million, has recently been built. Yet in spite of the huge effort that is needed, the problems of safety and land expropriation have given rise to rather strong unfavorable public opinion on the issue of water conservation.

TABLE 4.4

PUBLIC UTILITIES IN TAIPEI, 1950-1983

Year	Electric Power: K.W.H. Per Capita (Annual)	Electric Power: Percent of Capita Served	Piped Water: Liter Per Capita (Daily)	Piped Water: Percent of Capita Served	Telephones Per 1,000 Population
1950	316	43	67		10
1955	383	52	56		16
1960	513	61	84		23
1965	679	69	93	70*	34
1970	1,278	78	107	75	71
1975	1,351	93	117	91	148
1977	1,505	97	134	91	193
1983	1,630	99	387	99	335

*Figure for 1967.

Source: Taipei City Government, The Statistical Abstract of Taipei City, 1980.

Although programs for the construction of public facilities have absorbed more than sixty percent of the annual city budget in the past two decades, the standards have not far exceeded the minimum requirements. Parks and green fields in the city are few -- in 1977 Taipei had only 0.65 square meters of park space per person, in contrast to 19 square meters in New York. Schools, markets, public buildings, parking spaces and roads are all congested. Since the urban area is already densely populated, the construction of urban infrastructure confronts both sociopolitical difficulties and heavy financial burdens. Up to 1977, only a half of the planned area had been developed. Of the 1,128 hectares designed for parks and green fields, only twelve percent was developed. To complete the whole program, total estimated costs, not including the value of publicly owned land, are NT $48,042 million which is equivalent to two and a half times the total expenditures of the city government in 1977. Of the total costs, one half is for construction and half is for compensation for private land expropriation. Obviously, the scarcity of land is one of the critical problems in urban development.

Regulations have been approved which allow the land planned for specific uses to be used for multiple purposes on the condition that the original requirements are met, e.g., underground parking lots can be built under parks or school playgrounds; market places can be built with multi-stories for offices, urban services and low and medium income housing. This is surely an economical and efficient way of using scarce resources. But even if it is technically feasible, the costs are tremendously high compared with the costs of constructing the same infrastructure outside Taipei city.

Environmental pollution has also become one of the city's critical problems. Air pollution has to some extent been controlled in recent years, but it still remains at a level harmful to health. For example, dustfall was 22.5 metric tons per square meter per month in 1970; 21.9 tons in 1976, and then it dropped to 15.7 in 1983 (Bureau of Budget, Accounting and Statistics of Taipei City, 1983). Water pollution caused by human and industrial wastes is also serious. A modern sanitary sewer system is now under construction. The whole system is expected to be completed by the year 2020. Upon completion the system will be able to handle both human and industrial wastes for an estimated population of 6,270,000. As the project is very expensive and time consuming, it is not possible to wait for its completion to work out solutions to current problems. Rather, in the meantime it is urgent to enhance the capacity of the garbage disposal system, to enact a national anti-pollution law, and to educate public workers to carry out the laws and inhabitants to observe them.

In sum, recent rapid urbanization in Taipei has created a tremendous and urgent demand for urban infrastructure. This demand is met by diverting limited resources from alternative productive uses at unnecessarily high costs.

ECONOMIC GROWTH AND EMPLOYMENT

It has been indicated that the expansion of the population of young productive ages is of much greater magnitude than that of children and aged persons. It is also revealed by the 1975 agricultural census that among 264,192 persons who left their farms during 1973-1975, most came from central western rural areas, and 46 percent moved to Taipei city for nonagricultural employment. The 1980 Census showed a similar picture: of the 544,000 persons who moved across city or county boundaries between 1975 and 1980, 368,000 went to Taipei. The large influx of predominantly rural young migrants to the city has undoubtedly intensified the problem of employment, which the high rate of natural population increase has already made acute. This problem is indivisibly linked to problems of economic development of Taiwan as a whole. No intelligent solution to both can be achieved unless their nature and interrelationships are clearly understood.

It is a well known fact that since World War II Taiwan has possessed relatively vast reserves of manpower resources but has been short of capital and natural resources to develop them. Until 1952, 61 percent of a labor force of 2,936,000 were employed in agriculture, but the cultivated landholding per farm worker was only 0.49 hectares. Less than 10 percent of the labor force was employed in industries, which were dominated by traditional handicrafts. The final 30 percent was employed in the traditional service sector. Since labor was plentiful in both the agricultural and nonagricultural sectors, farming and industrial techniques were used that employed large numbers of workers at a low level of productivity. In order to improve this situation, the government adopted the goal of encouraging more intensive use of land in order to increase the agricultural surplus to support the growth of modern industries. This goal was satisfactorily achieved, on the one hand, by successful implementation of the "Land to the Tiller Program" in 1953, and, on the other, by measures transforming traditional industry from import to export substitution in the late 1950s and early 1960s (Galenson, 1979). As a result, the labor force engaged in agriculture decreased to 19 percent by 1983, while that engaged in industry increased to 41 percent. The real gross national product increased from NT $90 billion in 1952 to 1,173 billion (in 1976 prices) in 1983, or more than 13 times within a quarter of a century. The share of agricultural net domestic product decreased from 36 percent to 8.7 percent, while that of industry increased from 18 percent to 44 percent during the same period. These major structural changes in the labor force and the economy are reflected in two distinctive characteristics of Taiwan's development, namely, the attainment of full employment and the balance between urban and rural development:

1. *In spite of the rapid rate of labor force growth, recent economic progress has led to a marked improvement in employment in both urban and rural Taiwan.* Table 4.5 shows that the activity rates have remained stable at about 50 percent of the population aged 15 and over in Taipei city, and 60 percent in the rest of Taiwan. The unemployment rate for Taipei city has been significantly higher than that for Taiwan province over the period 1970-1984, but neither has reached a level considered serious. Underemployment rates, defined as percentage of the employed persons who work less than 36 hours per week to employed persons, were fairly high in the nonagricultural sector in earlier years in Taiwan as a whole. The situation has improved, however, as a result of successful implementation of land reform and industrial development. Toward the end of the period of observation the underemployment problem in quantitative terms virtually disappeared.

2. *The rapid growth and decentralization of industry have made it possible to achieve a balanced urban and rural development which, in turn, has prompted the utilization of labor resources that might otherwise not have been available.* Despite the rapid growth of Taipei in the 1960s and 1970s the distribution of employment by sector remained relatively constant in the city with the percentage employed in the industrial sector ranging from 32 percent to 38 percent (see Table 4.6). In contrast, the proportion

TABLE 4.5

LABOR FORCE PARTICIPATION, UNEMPLOYMENT, AND UNDEREMPLOYMENT IN TAIPEI CITY AND TAIWAN PROVINCE

	Average Annual Rates					
	Taipei City			Taiwan Province		
Years	Labor Force Participation	Unemployment Rate	Underemployment Rate	Labor Force Participation	Unemployment Rate	Underemployment Rate
1970-74	51.1	3.0	5.8	60.4	1.4	2.4
1975-79	50.8	2.8	3.9	61.2	1.7	2.8
1980-84	53.0	2.9	4.6	61.1	2.0	4.1

The labor force participation rate is defined as the percentage of the population aged 15 and over that is included in the labor force; the unemployment rate is defined as the percentage of the labor force that is unemployed; and the underemployment rate is defined as the percentage of the employed population that worked less than 36 hours per week. Beginning in 1978, the underemployment rate for Taipei City also includes Kaohsing City and the rate for Taiwan Province excludes Kaohsiung.

Source: Quarterly Report on the Labor Force Survey in Taiwan, prepared by Taiwan Provincial Labor Force Survey and Research Institute, for respective years in July.

employed in industry in the rest of Taiwan increased from 21 percent in 1966 to 43 percent in 1984. Since 1975, the proportion of the labor force employed in industry has been greater outside the city than within the city. These facts imply that to some extent the government has succeeded in taking measures to discourage the concentration and development of industries in the crowded Taipei city proper. However, the rapid concentration of industries in the areas immediately surrounding the city is still a problem requiring serious consideration. The relatively large increase in service employment in Taipei city was surely a response to the needs for modern services required by the rapid expansion of manufacturing employment throughout the island.

Obviously, the problems of rapid growth of the labor force and accompanying unemployment, which are witnessed in most less developed countries, do not appear to be problems in Taiwan. However, the low unemployment rate and the rapid expansion of labor-intensive industries serve to mask what is a more serious problem, that of inadequate employment or labor underutilization.

These findings make it clear that there is no problem of supplying jobs to support the influx of rural migrants to Taipei. The problem rather, lies in the provision of adequate jobs with respect to income and education levels. The basic causes of inadequate labor force utilization in Taiwan are closely linked to demographic variables and traditional values. As discussed before, the post-World War II population explosion, which stemmed from a rapid reduction in mortality not offset by a decline in fertility, laid an immediate economic burden on the war-devastated nation and an enormous demand for new jobs in later years. To cope with these problems in line with its principle of equal opportunity for all, the government at that time had to pursue a policy of developing labor-intensive techniques in both agricultural and nonagricultural sectors.

This policy undoubtedly released the population pressure on resources and employment, but it could not have created a linkage to modern economic growth if appropriate measures had not been taken in time to transform the structure of the economy from a labor-intensive one to capital - or technology-intensive one. One reason for this is that, by tradition, Chinese are education-oriented. The majority of parents intend to drive their children to achieve the highest possible level of education irrespective of the needs of the economy. Consequently, many white collar jobs have been filled with overqualified persons whose education and training are not suitable to their positions. This has made them displeased not only with their salaries but also with their work. Ironically, at the same time, the blue collar jobs have been unable to attract enough well-trained skilled workers, or even stable unskilled workers. As a result, the wage rate for blue collar jobs has been low, which has further aggravated the problem of finding sufficient workers to fill the jobs.

The birth rate for Taipei city declined from 41 births per thousand population in the early 1950's to 17 in 1984. Further decline is expected to depend on changes in attitude toward family size. It is observed that the

TABLE 4.6

DISTRIBUTION OF EMPLOYMENT IN TAIPEI CITY AND TAIPEI PROVINCE 1966-1984

	Taipei City				Taiwan Province			
	Agriculture	Industry	Services	Total	Agriculture	Industry	Services	Total
				Numbers				
1966	5	114	194	313	1783	758	1057	3598
1970	17	196	304	517	1787	1099	1311	4197
1975	17	202	397	616	1751	1742	1531	5024
1980	31	411	708	1149	1377	2373	1818	5568
1984	35	457	810	1303	1386	2636	2091	6113
				Percentages				
1966	1.6	36.4	62.0	100.0	49.5	21.1	29.4	100.0
1970	3.3	37.9	58.8	100.0	42.6	26.5	31.2	100.0
1975	2.8	32.8	64.5	100.0	34.9	34.7	30.5	100.0
1980	2.7	35.8	61.6	100.0	24.7	42.6	32.7	100.0
1984	2.7	35.1	62.2	100.0	22.7	43.1	34.2	100.0

Note 1: From 1966 to 1975 Taiwan Province includes the entire island except for Taipei City. From 1980 and 1984, Kaohsiung City is also excluded from Taiwan.

2: Agriculture includes agriculture, forestry and fishing; industry includes mining, manufacturing, construction and utilities; services include all other civilian jobs.

Source: Quarterly Report on the Labor Force Survey in Taiwan.

previous reduction in birth rate has been achieved by eliminating excess births to reach a rather large ideal family size: three to four children with at least two boys. This implies that the government's population policy in the immediate future should place priority on changing attitudes regarding the ideal family size and the preference for male children, rather than on the provision of knowledge and means of birth control, which existing family planning programs stress. The shift in emphasis of the family planning program is particularly important for Taipei city, because the bulk of new migrants to Taipei are of reproductive age and rural background.

The problem of inadequate employment brought on by the rapid population growth in the postwar period has been only gradually recognized as a major issue in recent years. Policies to deal with this problem have been neither adequately designed nor effectively implemented on a sufficiently large scale. For example, in Taipei in 1984 only 10 percent of the unemployed persons used government employment agencies to obtain job information. Thirty-five percent obtained job information from relatives or friends, 5 percent used private employment services, while 46 percent relied on newspapers for job information (Taiwan Provincial Labor Force Survey, 1984). Even this low level of government employment services has been made available only in recent years in Taipei.

The government's educational programs do not seem to be sufficiently adjusted to cope with the problem of labor underutilization. Changes in the composition of education attainment of the labor force by occupation in recent years, as shown in Table 4.7, clearly indicate that, with the exception of the insignificant number of farmers in Taipei city, the educational level of the labor force in both Taipei city and Taiwan province has been improved. However, this improvement is mainly concentrated in general education. Improvement in skill training, particularly among the blue collar workers, is still greatly needed.

Looking at the changes in Taiwan's economic structure, the picture is less discouraging. In fact, it is likely that labor market force could be an important mechanism for adjusting the inadequate labor utilization. Table 4.8 shows that the average real income per household of workers in Taipei city increased 66 percent during the period 1971-1983, although there was a violent dip in 1974 due to the oil crisis. The increase in real income, however, is not evenly spread among occupation groups. There is an obvious positive correlation between the degree of inadequate utilization by income level and the rate of income increase. This finding perhaps implies that, excluding the influence of uncontrollable economic disturbances, there are built-in factors in the economy, attributable to structural changes, which promote better labor utilization. This is also apparent when we look at the discrepancies between Taipei city and Taiwan province in per household income of workers. The ratio of average real income per household by occupation in Taipei to that in Taiwan over the period 1971-1983, as shown in Table 4.8, indicates, first, that the income gap between these two areas is narrowing, and second, that the degree of inadequate utilization by income level tends to be positively associated with

TABLE 4.7

EDUCATIONAL ATTAINMENT OF LABOR FORCE BY OCCUPATION IN TAIWAN, 1970-1984

	Taipei City					Taiwan Province				
Year	College and Higher	Senior Voca-tional School	High School	Primary and Lower	Total	College and Higher	Senior Voca-tional School	High School	Primary and Lower	Total
White-Collar*										
1970	38.4	28.0	24.4	9.1	100.0	17.6	33.3	31.2	18.0	100.0
1975	43.3	21.9	24.0	10.7	100.0	29.0	31.9	25.9	13.2	100.0
1980	47.3	24.8	20.5	7.3	100.0	37.4	28.5	22.7	11.3	100.0
1984	50.0	25.7	19.2	5.1	100.0	40.6	31.1	19.8	8.5	100.0
Blue-collar*										
1970	2.1	4.3	18.6	75.0	100.0	0.5	4.0	16.5	79.0	100.0
1975	2.7	7.6	22.7	67.0	100.0	1.0	7.1	24.9	67.0	100.0
1980	3.8	12.1	31.0	53.1	100.0	2.0	10.2	33.1	54.7	100.0
1984	4.6	14.2	33.6	47.7	100.0	2.7	15.6	35.0	46.6	100.0
Sales & Service Workers										
1970	3.0	6.7	23.6	66.7	100.0	1.5	6.3	19.8	72.3	100.0
1975	7.1	9.9	26.4	56.6	100.0	2.9	8.4	22.1	66.6	100.0
1980	14.4	11.4	28.7	45.5	100.0	5.8	11.5	25.9	56.8	100.0
1984	13.2	14.7	27.9	44.2	100.0	7.8	16.3	27.8	48.1	100.0

*White-collar category includes professional, technical, administrative, managerial, and clerical workers. Blue-collar category includes miners, quarrymen, transport workers, craftsmen, and production- process workers.

Source: Quarterly Report on the Labor Force Survey in Taiwan, prepared by Taiwan Province Labor Force Survey and Research Institute, for respective years in July.

TABLE 4.8

AVERAGE REAL INCOME PER HOUSEHOLD BY OCCUPATION FOR TAIPEI CITY AND RATIO OF TAIWAN PROVINCE TO TAIPEI CITY 1971-1984

	Profes- sional Technical	Adminis- trating Managerial	Clerical	Sales	Service	Blue- Collar Workers	All Occu- pations
	Average Real Income in Constant 1971 Dollars (1000 NT$)						
1971	91	124	79	69	52	56	74
1978	118	159	110	103	90	84	102
1983	146	168	128	126	102	96	123
Increase 1971 to 1983	60%	36%	61%	83%	97%	71%	66%
	Ratio of Income in Rest of Taiwan to Taipei City						
1971	.68	.68	.72	.79	.78	.74	.63
1978	.80	.77	.76	.73	.71	.75	.68
1983	.88	.94	.85	.71	.80	.84	.70

Source: Report on the Survey of Family Income and Expenditure, prepared by Taipei City and Taiwan Provincial Governments.

the ratio of income discrepancies. Free labor mobility among occupations and between rural and urban areas is evidently one of the most important factors contributing to this phenomenon. The government's strong emphasis on rural development and decentralization of industry is also of primary importance. This again illustrates that urban problems should be tackled as part of an integrated nationwide plan.

CONCLUSION

This chapter has traced the growth of Taipei from its origin in the 17th century to the mid 1980s. Taipei has been the capital of Taiwan for most of the last century and as such has experienced rapid growth. Since 1950, most of the growth of Taipei has been outward expansion into the more peripheral parts of the city and beyond the boundaries of the city. This resulted in the annexation of six towns in 1968 and the rapid growth to separate city status of several surrounding towns. Since 1970, growth in the old districts of the city has come to a standstill and the central business district has actually lost population.

Rapid urbanization in the Taipei area is mainly due to the continuing large influx of migrants from the rest of the island in response to both demographic and economic factors connected with population pressure and to rapid expansion of job opportunities in Taipei. The fairly balanced growth between the agricultural and industrial sectors has alleviated some of the economic and employment problems, but many still remain to be solved. Most of the migrants are engaged in labor-intensive manufacturing or service jobs. The new migrants do not have difficulties in finding jobs in Taipei, but the jobs they find are not adequate in terms of income and utilization of the education of the migrants. General economic policies have helped bring about a more adequate utilization of the labor force in recent years. However, it is essential that the government continue to stimulate the growth of higher productivity jobs which can offer continued growth and income while taking advantage of the rising levels of education of the work force.

The growth of the city has not been planned very efficiently. *Ad hoc* measures for the development of various urban services and facilities, implemented primarily to keep pace with the increase in population, have alleviated and to some extent improved the situation. The quality and scope of such measures however, should be greatly improved, and more attention and greater material and technical resources should be devoted to their implementation.

Because of the physical constraints on growth in the Taipei basin, it is important that many of the new job opportunities be located away from Taipei. In the next chapter we shall look at the growth of industrialization in rural areas and commuting of workers who remain in rural areas as alternatives to rural to urban migration.

5

Alternate Responses to a Changing Economy: Migration, Commuting and Rural Industrialization

In the three decades from the early 1950s to the early 1980s, Taiwan underwent a rapid transition from an agrarian based economy to an industrial economy. In 1951, 56 percent of the labor force was engaged in primary sector activities, principally small scale farming, and only 17 percent of the labor force was employed in the secondary sector (manufacturing, mining, construction or utilities). By 1980 the proportion employed in farming and the primary sector activities had dropped to less than 20 percent and the proportions employed in manufacturing and other secondary sector activities had risen to 42 percent (see Table 5.1). Between 1980 and 1985, the proportion employed in agriculture declined further to 17.5 percent, with most of the gains in the tertiary or service sector.

This transformation was accompanied by urbanization, but the movement of people from rural to urban areas was nowhere near as great nor as disruptive as it could have been. While the proportion of the labor force employed in agriculture declined, the sector grew in absolute numbers until 1969 and by 1985 it had declined by about 25 percent. Much of this decline has been due to retirement; few established farmers have left the land for factory work.

What has happened is that the majority of young adults entering the labor force have gone into manufacturing or other nonfarm jobs. This is clearly illustrated by the age structure of employees in manufacturing in comparison to farmers and other primary sector workers. As of 1973, those in manufacturing tended to be young with a median age of 28 for males and 19 for females. In contrast, males in the primary sector had a median age of 38 and females a median age of 34. While many of those entering manufacturing have moved to the cities, many others have been able to find nonfarm work nearer to thier homes due to the dispersion of industrialization throughout urban and rural areas of Taiwan. Others have been able to commute to work from rural residences.

TABLE 5.1

EMPLOYMENT BY SECTOR (PERCENTAGE DISTRIBUTION)

	Primary	Secondary	Tertiary	Total Labor Force	
				Percent	Number
1952	56.1	16.9	27.0	100.0	2929
1955	53.6	18.0	28.4	100.0	3108
1960	50.2	20.5	29.3	100.0	3473
1965	46.5	22.3	31.2	100.0	3763
1970	36.7	28.0	35.3	100.0	4576
1975	30.4	34.9	34.7	100.0	5521
1980	19.5	42.4	38.1	100.0	6547
1985	17.5	41.4	41.1	100.0	7428

Source: Taiwan Statistical Data Book 1986, p. 16.

MIGRATION BETWEEN CITIES AND COUNTIES

The in, out, and net movement for each country and major city has been recorded in the household register since 1951 and was also tabulated in the 1980 census. While migration registration is good, it is not complete. Many moves, especially those of short duration, are not registered, some registered moves are changes in legal place of residence not accompanied by movement, and there are substantial delays in registration (Speare, 1971b; Speare, et al., 1975). Despite these difficulties, registered migration provides a good estimate of the relative size of population movements and many errors cancel out when net migration is calculated.[1]

The pattern of net migration shown in Table 5.2 is one of rural to urban movement. Throughout the 30 year period, almost all of the net

---

[1] Since the mid-1970s, the population register in Taipei City and some of the other large cities has had difficulty in keeping up with the registration of in-migrants and the accuracy of registration data for measuring net migration to these cities has become questionable. For that reason, we have used the 1980 Census to estimate net migration from 1975 to 1980.

migration has gone to the five major cities or to Taipei County which surrounds Taipei City. Taipei City and the satellite cities in Taipei County in the north, and Kaohsiung City in the South have received the largest numbers of migrants. In recent years, the volume of migration has been greater to Taipei County than to the city itself. Taichung City increased substantially in net migration from the first period to the second, making it the third major center of attraction for migrants from 1963-74. However from 1975-80 it received few migrants. At no time during the 30 year period was there very large migration to the other two cities, Keelung and Tainan, and for the last period Keelung had net out-migration. While both of these cities played important roles in the prewar development of Taiwan, their functions have been taken over by other cities.

The growth in Taipei County and Taoyuan County is primarily the result of the expansion of Taipei City and the growth of satellite centers along the major transportation lines emanating from the city from which people can commute daily to Taipei city. In the latest period, these satellite cities have become major sites for the location of new industries as well. The intermediate sized cities along the main railway line in Taoyuan County, Taoyuan and Chungli, had net migration around 4 percent per year during 1963-74 despite the fact that they are located more than one hour from Taipei by train.

In the period from 1975-80, there was also net migration into suburbs and satellite cities surrounding Taichung City and Kachsiung City, as shown by the positive net migration rates for the two counties with the same name as these cities. Apart from these counties, the only other counties with positive net migration for any of these three periods were Taitung and Hualien on the East Coast which gained during 1951-62 because of improved methods for farming hilly areas and the construction of the cross-island highway linking these areas to the Western part of Taiwan.

Overall, the data on net migration show a continued flow from most of the counties to the major cities and their surrounding suburbs and satellite cities. In most cases these flows increased significantly between the first 12 year period which roughly correspond to the takeoff phase of Taiwan's development and the second 12 year period which corresponded to the labor-intensive phase of development. Since 1975, migration to the major cities has slowed considerably, although net migration out of most counties has continued at levels comparable to those of the labor-intensive period, when adjustment is made for the shorter time period. However, much of this migration appears to have gone into suburbs and satellite cities located outside the boundaries of the major cities.

While net migration was clearly in the direction of the major cities and their satellites during the entire 30 year period, the rates of movement were modes when the population size in place of origin is considered. For example, the 1,025,100 migrants who left the 14 countries with net out migration during the labor-intensive period of development from 1963-74 constituted less than one percent of the population per year. Over the

TABLE 5.2

NET MIGRATION FOR MAJOR CITIES AND COUNTIES, 1951 to 1980 (in thousands)

	1951-62	1963-74	1975-80
North Region	267.5	563.8	381.0
Taipei City	177.8	290.1	38.5
Keelung City	14.3	2.1	-21.1
Taipei	114.3	330.6	351.0
Taoyuan	0.8	44.4	71.1
Hsinchu	-23.5	-53.2	-33.8
Ilan	-16.2	-50.2	-24.8
Central Region	-231.6	-368.4	-183.7
Taichung City	16.1	80.1	6.1
Miaoli	-42.2	-69.2	-41.4
Taichung	-48.3	-45.4	32.6
Changwa	-88.2	-131.4	-63.8
Nantou	-13.6	-61.9	-36.9
Yunlin	-55.4	-140.6	-80.3
South Region	-119.9	-186.1	-154.3
Kaohsiung City	66.6	200.0	34.0
Tainan City	26.1	22.7	3.6
Chiayi	-57.5	-149.6	-81.7
Tainan	-85.0	-145.2	-61.1
Kaohsiung	-29.9	-5.0	9.6
Pingtung	-24.7	-77.6	-46.0
Penghu	-15.5	-31.4	-12.8
East Region	42.3	-64.4	-43.0
Hualien	3.0	-27.4	-15.6
Taitung	39.3	-37.0	-27.4

Sources: 1951 to 1974: Taiwan Demographic Fact Book; 1975-80, 1980 Census.

Notes: Yangmingshan is included with Taipei County from 1951 to 1958 and with Taipei City from 1959. From 1967, Taipei City includes 4 townships which were annexed from Taipei County in 1968.

entire 30 year period, the movement amounted to 1,966,200 persons or 17 percent of the 1980 population. Given the decline in agriculture from 59.3 percent to 19.5 percent during this period, the migration could have been much larger. However the volume of migration was kept down by the development of industrial employment in rural areas and the ability of many rural residents to commute to nonfarm jobs.

RURAL INDUSTRIALIZATION

Taiwan's urbanization would have been much more rapid had it not been for the spread of industrialization into rural areas. Had all of the industrialization taken place within the cities, growth rates of these cities might have exceeded 6 percent during the 1960s and 1970s.[2] However, a significant amount of industrialization occurred outside cities in small towns and rural areas.

The origins of rural industry can be traced to the development of agroindustries during the 1930s under the Japanese (Ho, 1979). One of the first major rural industries was sugar refining which was followed by canning of fruits and vegetables. The development of these rural industries and the ones to follow was promoted by investment in infrastructure. During the 1930s, the Japanese built schools in rural areas and sought to make primary school education universal. They also began the process of rural electrification, which by 1960, included 70 percent of rural households (Ho, 1979). In addition, the Japanese began the development of an extensive system of rural roads. By 1970, Taiwan had about 215 kilometers of paved roads per 1000 square kilometers in contrast to only 50 kilometers per square kilometers in Korea in 1975 (Ho, 1979).

Land reform undoubtedly played an important role in facilitating rural industrialization in Taiwan. By granting ownership to the tenants of the small plots which they were tilling, land reform greatly increased the number of households which had a stake in remaining in rural areas. Because these plots were typically too small to provide full employment for all children, rural households typically had extra labor as their children reached the normal age for labor force entry. This meant that there was a supply of labor in rural areas which had an attachment to these areas and thus was more willing to work in or near the village than to move to the city.

Rural industrialization was also encouraged by various government

----

[2] Between 1960 and 1980, manufacturing employment grew at an average annual rate of over seven percent. If most of this manufacturing were located within the boundaries of cities, the cities might have also grown at this rate rather than the four to six percent rates shown in chapter 2.

policies. Between 1968 and 1981, the government, in conjunction with private corporations, established 62 new industrial zones in Taiwan. Only 10 of these were in one of the 5 large cities, 22 were in one of the four "metropolitan counties" surrounding these cities, and 30 were in the remaining more rural counties (Tsai, 1982, p. 7).

Finally, another factor promoting rural industrialization was agricultural development which lead to increased farm output through the use of improved seeds, fertilizer, irrigation, etc. This increased productivity resulted in growing farm incomes and a growing demand for nonfood consumption (Ho, 1979). For example, there was an increased demand for building materials such as bricks and roof tiles which could easily be produced by small factories in rural areas. In addition, there was a growing demand for clothing which was met, in part, by an expansion of textiles and apparel manufacturing in rural areas.

Selya (1974) describes the outward expansion of industrial sites as a "leap frog" pattern. "Plants are strung out with major interruptions in the pattern by paddy or marginal land. These intervening spaces may be filled in later if and when the farmer owning the land decides to sell. The landscape then assumes a mixed appearance, even within the industrial sectors: old and new often stand side by side (Selya, 1974, p. 55).

The growth of manufacturing in rural areas is reflected in the official figures for the number of factories by county. In 1979, 20,996 out of 58,465 factories, or 36 percent, were located in one of the 12 more rural counties (Tsai, 1982, p. 12).

The rural areas tended to attract the more labor-intensive industries. Ranis (1979, pp. 229-232) estimated that the ratio of total assets per worker in manufacturing was about twice as high in urban areas as it was in rural areas in 1971. Ranis points out that industries specializing in producer goods were much more likely to be located in one of the five large cities than industries specializing in consumer goods. In addition to food processing and lumber where the attraction of rural areas is obvious, textiles, chemicals and metal manufacturing were attracted to rural areas (Tsai, 1982). This relationship between capital intensity and level of urbanization of manufacturing locations was further strengthened by the national development plan in 1979 which stated that labor-intensive industries should go to small towns and rural areas (Tsai, 1982).

OFF-FARM WORK IN RURAL HOUSEHOLDS

While rural industrialization provided opportunities for non-farm work in rural areas, the expansion of highways and public transportation made it easier for rural residents to commute to jobs in urban areas. The increase in off-farm work can be clearly seen in a series of three studies of rural farm households which are near major cities Tsui and Lin (1964), Lin and Chen (1969) and Liao (1976). These studies have documented not only a large increase in off-farm work but have also shown that much of this involves commuting and seasonal work rather than long-term migration. Between 1963 and 1975, the number of off-farm workers per 100 farm households increased from 137 to 198. According to Chinn (1979) the proportion of household income for farm females which came from nonfarm work increased from 11 percent in 1960-62 to 31 percent in 1970-72.

They classified the off-farm workers into three categories:

1. *Commuters* who travel regularly back and forth from home to work and receive a monthly salary.

2. *Seasonal workers* who work temporarily for others and receive wages per working day.

3. *Long-term migrants* who leave home and work rather permanently in the city or some other place but retain close connections with the farm home.

While these definitions leave ambiguous how such groups as commuters who are paid by the day should be classified, they do make the important distinction between those who have moved out and those who continue to keep the farm house as their major residence.

In 1963, more than one-half of all off-farm workers were seasonal workers. Four out of five of these seasonal workers were employed in agriculture, but worked as temporary laborers on someone else's farm. There were 24 long-term migrants and 35 commuters per 100 farm families in 1963 (see Table 5.3). By 1975 there were 91 long-term migrants and 74 commuters, a substantial increase. However, by 1975, the number of seasonal workers had declined to 33 per 100 households and these constituted only 17 percent of the off-farm workers. The proportion of farm workers among this group had fallen to 53 percent in 1968 (Lin and Chen, 1969:15) and was probably lower in 1975, although Liao does not present data on the occupation of seasonal workers.

Males predominated among all types of off-farm workers during all three periods. However, the proportion of females increased from 21 percent in 1963 to 33 percent in 1975 due to greater involvement of females in commuting and long-term migration. This finding is consistent with the overall increase in female employment in nonfarm occupations in Taiwan which was documented in chapter 2. Most of the female migrants and commuters were young--about three-quarters were under age 25 indicating that such work was most common between the completion of school and marriage or during the early years of marriage. In contrast, males had a wide distribution of ages with commuters and seasonal migrants being older, on the average, than long-term migrants.

Off-farm employment was highly correlated with education (see Table 5.4) Among the working age population in the farm households surveyed, the proportion working outside the farm varied from 11 percent for those with no education to 91 percent for those with college education. In

TABLE 5.3

NUMBER OF OFF-FARM WORKERS PER 100 FARM HOUSEHOLDS

	1963[a]	1968[b]	1975[c]
Males			
Commuters	25	54	45
Seasonal Workers	66	33	28
Long-term Migrants	17	31	60
Sub-total	108	118	133
Females			
Commuters	10	23	29
Seasonal Workers	12	9	5
Long-term Migrants	7	10	31
Sub-total	29	42	65
Total Males and Females	137	160	198

[a]Calculated from Tsui and Lin (1964), Tables 7-9. These figures differ from those presented in Tsui and Lin (1964, Table 2) for long-term migrants. It appears that in calculating out-migrants per household, Tsui and Lin included migrants who were not in the labor force. Their rates for long-term migrants are 33 males and 10 females per 100 households.

[b]Lin and Chen (1969), Table 3.

[c]Liao (1976), Table 4.

TABLE 5.4

EDUCATION AND OFF-FARM WORK FOR WORKING AGE POPULATION (Aged 15 and over), 1968

	Commuters	Seasonal Workers	Out-Migrants	Remaining at Home	Total Percent	N
No Education	3.1	7.0	0.6	89.4	100.0	1825
Primary School	14.8	10.7	8.7	65.9	100.0	4701
Junior High School	35.7	4.6	18.8	41.0	100.0	591
Senior High School	50.0	2.1	23.5	24.3	100.0	480
College	46.7	0.0	44.4	8.9	100.0	45
Total	16.0	8.7	8.7	66.6	100.0	7642

Source: From Lin and Chen (1969), Table 10.

addition, those with no education or only primary school education were much more likely to be seasonal workers than those with higher levels of education.

The relationship between the type of off-farm work and education fits the household response model well. With the increasing pressure on the land and the opening up of alternative opportunities for employment, the type of response appears to have been largely determined by the human assets possessed by the household. Where household members had relatively high levels of education they sought employment outside the village where their skills could be best used. The higher their education, the more likely they were to make a long-term move, although at all levels of education, except the lowest, commuting was the preferred response. Those with intermediate levels of education, such as the junior high school graduates, were still more likely to work off the farm than on the farm and most of these chose commuting as their mode of off-farm work. Those with less than average education were much more likely to remain on the family farm and if they worked outside the farm, they were likely to choose seasonal work.

Unfortunately, none of these three studies report the location of the work place for household members working off the farm. Tsui and Lin (1964:16) present data on commuting time which show that the median commuting time was between 15 and 30 minutes in the North and South region. From these data we can conclude that much of the work was probably in the same township or at least within the same county.

STUDY OF LABOR MOBILITY 1967-72

A clearer picture of the labor force transformation can be obtained by looking at the changes experienced by a cross section of the entire labor force of Taiwan over a five year period when change was particularly rapid. This study is based on the analysis of a set of supplementary questions which were appended to the July 1972 labor force survey and which asked about employment status, place of work and place of residence five years prior to the survey.[3] From the cross tabulation of these items with the standard questions about current employment status, place of work and place of residence, we have been able to estimate the magnitude of changes which occurred between 1967 and 1972 and the mechanisms responsible for these changes.

---

[3] The Labor Force Survey was conducted by the Taiwan Provincial Labor Force Survey and Research Institute. Approximately 17,000 persons were included in the July 1972 survey and they were selected from 98 primary sample units. Results have been weighted to represent all Taiwan.

Industrial Mobility of Labor

Between 1967 and 1972, the number of civilian employed persons increased from about 4 million to 5 million. About 1.4 million persons entered employment whereas 0.3 million left employment and an estimated 95,000 persons who were employed in 1967 died.[4] In addition to the turnover in the employed population, an estimated 368,000 persons who were employed in both 1967 and 1972 changed jobs across major industrial categories. The magnitude and net effect of these two components of industrial mobility can be seen in Table 5.5.

The net rate of increase of the employed population was 22 percent of the 1972 employed population. This rate of increase varied from a negative rate for agriculture and mining to 49 percent for manufacturing. The growth rate for commerce, finance, and services was above average and the growth rate for utilities, construction, and transportation was about average.

For most of the growing industries, net entries into the industry represented a much larger component of growth than movement from other industries. For manufacturing, there were an estimated 526,000 net entries, but only 60,000 net movers from other industries. The situation was similar, but not as pronounced for construction, commerce, and finance. Services had a relatively large net entry of 203,000 persons, but a net out movement of 4,000 persons, indicating that some people may have found temporary first employment in services prior to obtaining more permanent employment in another industry.

The major shift in the industrial distribution was from agriculture to manufacturing. If agriculture had grown at the average rate for all industries, it would have had 2,157,000 persons in 1972 or 488,000 more than were employed in agriculture in 1972. Only a small part of the relative loss to agriculture was due to the mobility of persons who had been employed in agriculture in 1967. The rate of movement out of agriculture was only 9.1 percent for the five year period which was lower than the rates of movement out of mining, transportation, banking and insurance. While agriculture did have a distinctively low rate of entry from other industries, it suffered most from its failure to attract a proportionate share of persons entering the labor force during the period. While 43 percent of the employed population were in agriculture in 1967, only 19 percent of the

----

[4] The mortality estimate was obtained by applying life table survival rates to the age and sex distribution of the employed population in 1967 as reported in the results from the 1967 labor force survey. We have assumed that the survival rates are the same for the employed population and the total population.

TABLE 5.5

COMPONENTS OF CHANGE IN THE LABOR FORCE BY INDUSTRY, TAIWAN, 1967-72
(thousands of persons)

			Inter-Industry Mobility				Entries and Exits from Employment		
Industry	Number Employed in 1972	Survivors by 1967 Industry	Net Change 1967 to 1972	Net Mobility	Moved in from Other Industries	Moved Out to Other Industries	Net Entries	Entries	Exits
Total	4972	3866	1107	0	368	368	1105	1433	326
Agriculture	1669	1677	-8	-126	25	151	117	268	150
Mining	73	75	-2	-2	8	10	0	6	7
Manufacturing	1185	599	587	60	119	58	526	581	55
Utilities	36	28	8	3	3	0	5	6	1
Construction	265	204	62	16	38	21	45	57	11
Commerce	664	481	182	39	82	43	143	179	36
Transportation	247	192	55	11	39	28	44	56	12
Banking and Insurance	94	70	24	1	9	9	23	29	6
Services	740	541	199	-4	44	48	203	251	48
Percentage of Number Employed in 1972									
Total	100.0	77.8	22.3	0	7.4	7.4	22.2	28.8	6.6
Agriculture	100.0	100.5	-0.5	-7.6	1.5	9.1	7.1	16.1	9.0
Mining	100.0	102.7	-2.7	-2.7	10.0	13.7	0	8.2	9.6
Manufacturing	100.0	50.6	49.5	5.1	10.0	4.9	44.4	49.0	4.6
Utilities	100.0	77.8	22.2	8.3	8.3	0	13.9	16.7	2.8
Construction	100.0	77.0	23.4	6.0	14.3	7.9	17.0	21.5	4.2
Commerce	100.0	72.4	27.4	5.9	12.4	6.5	21.5	27.0	5.4
Transportation	100.0	77.7	22.3	4.5	15.8	11.3	17.8	22.7	4.9
Banking and Insurance	100.0	74.5	25.5	1.1	9.6	9.6	24.5	30.9	6.4
Services	100.0	73.1	26.9	-0.5	6.0	6.5	27.4	33.9	6.5

entrants went into agriculture.

The shift in the industrial composition of the labor force was largely accomplished through the new entrants to the labor force who chose a mix of industries considerably different from that of the existing labor force. This shift is not an artifact of higher turnover rates in some industries. Were this the case, there would be a strong relation between entries and exits. However, manufacturing, which has the highest rate of entry, has one of the lowest rates of exit.

The Geographical Mobility of Labor

The Labor Force Study data include a tabulation of the county or major city of residence in 1967 and the county or major city of residence in 1972. The five largest cities, which comprised 73 percent of the population of places over 100,000 in 1970 can be separately identified. Because the sample is not necessarily representative for each city or county, we have grouped the data into two city groups and four regions.

The results, which are shown in Table 5.6 indicate a net movement of the working age population to the major cities and the Northern region and away from the other regions. The volume of movement of persons employed in both 1967 and 1972 is relatively low and most of the redistribution of workers appears to be due to entries and exists from the employed population. Although Taipei City had a lower rate of net entry than most of the other areas, this was due primarily to the low rate of natural increase for the city. Additional tabulations, which are not shown here, indicate that there was a net movement of about 13,000 new entries to Taipei City.

The largest net change occurred in the Northern region which includes the townships surrounding Taipei City. Mobility data from the household register show that most of the net migration into the Northern region has gone to four townships adjacent to Taipei City. In 1971, the net migration to these four townships was 27,100 which is greater than the net migration to the entire region.[5]

These results are consistent with those from the household register with respect to the direction of the net flows but not the volume of these flows. For example, the labor force survey shows less than one percent net migration of workers to Taipei City and only 2.2 percent net migration of workers to other large cities. The household register, which provides data only for the total population, shows a net migration of 7.6 percent to Taipei

----

[5] It is possible for the net migration to a few towns to exceed the net migration for the region if other towns in the region had net out-migration. These data were derived from the *1971 Taiwan Demographic Fact Book*, Ministry of Interior, Republic of China, November, 1972.

TABLE 5.6

COMPONENTS OF CHANGE IN EMPLOYED LABOR FORCE BY RESIDENTS

	Employed Labor Force 1972	Survivors of 1967 Labor Force by 1967 Residence	Net Change	Mobility of Persons Employed at Both Times			Change due to Entries to and Exits from Employment: Net Entries	Entries	Exit
(Numbers in thousands)									
Total	4965	3865	1100	0	119	119	1100	1426	326
Taipei City	569	454	114	1	28	27	113	156	42
Other Cities	647	485	163	14	32	18	148	190	42
North	1287	948	339	9	30	21	330	396	68
Central	1474	1172	302	-14	16	30	317	415	98
South	744	602	143	-8	10	18	150	207	57
East	244	204	39	-3	2	5	42	61	19
(Percentage Distribution)									
Total	100.0	77.8	22.2	0.0	2.4	2.4	22.2	28.7	6.6
Taipei City	100.0	79.9	20.1	0.2	4.9	4.7	19.9	27.4	7.5
Other Cities	100.0	74.9	25.1	2.2	5.0	2.8	22.9	29.4	6.5
North	100.0	73.7	26.4	0.7	2.3	1.6	24.6	30.9	5.3
Central	100.0	79.5	20.5	-1.0	1.1	2.1	21.5	28.1	6.7
South	100.0	80.9	19.1	-1.1	1.4	2.4	20.2	27.8	7.6
East	100.0	83.8	16.2	-1.2	0.8	1.9	17.3	25.0	7.7

City and 8.2 percent to the other large cities for a comparable five year period from the end of 1966 to the end of 1971. A large part of the difference can be attributed to biases in the regional distribution of the sample for the labor force survey. Although these biases are not large enough to have much effect on the industrial distribution of workers, their effects tend to be magnified in estimating net migration because the in-migrant and out-migrant estimates are obtained from different parts of the sample.

Another way of studying the geographical mobility of labor is to look at changes in the place of work. As was the case for change in place of residence, employed labor force was greatest in the cities and the Northern region. Similarly, only a small proportion of the total change in the employed labor force was due to persons who were employed in 1967 and 1972 and who changed their place of work. The largest component of change was due to net entries into employment. The rate of entry into employment was about twice as large for the cities and the Northern region as it was for the other regions. This means that most of the geographical mobility of labor occurred at or near the time of entry into employment.

Taipei City had a net increase of about 270,000 employed persons which was due to a net mobility of 29,000 and net entries of 241,000. The net entries were the difference between 295,000 persons who entered employment and 54,000 who left employment. Since the proportion leaving employment in Taipei was similar to that for all of Taiwan, little change in the distribution of places of work resulted from people leaving employment. The mobility of people entering employment can be estimated by comparing the expected number of new entries to the actual number. On the basis of the age distribution in Taipei in 1967, we have estimated that the city should have had about 150,000 new entries from its own population. This means that about half of the people who entered employment in Taipei came from other areas. This is considerably larger than the estimated 29,000 workers who had been employed outside Taipei in 1967 who changed their place of work to Taipei. However, not all of those who worked in Taipei, lived in Taipei. Many lived outside the city and commuted to work.

Commuting to Work

Only 63 percent of those who worked in Taipei City had their legal residence there. Most of the others lived outside the city and commuted daily to work there. However, some workers, mostly recent migrants who had left their family behind, stayed at the temporary residence in the city and did not change their legal residence (see Speare, 1971b and Speare, et al., 1975). Most of those who worked in Taipei City but lived outside the city, lived in the Northern region. However, 14 percent of those who worked in Taipei had their permanent residence in the Central, South, or East region from which daily commuting would have been difficult.

Presumably these workers had temporary residences in Taipei or the area surrounding the city.

A higher percentage of the people who worked in other cities also had their legal residence in these cities. The average was 79 percent. For the regions, the proportion who lived in the same region where they worked was even higher. Although we do not know the extent of commuting within regions, we can conclude that Taipei City had the largest proportion of commuters and temporary residents and that it drew them from greater distances than other areas.

Further evidence on the importance of commuting comes from a special study of workers in the Kaohsiung and Nantze Processing Zones carried out by Rong-i Wu. He found that about one-quarter of the workers in these zones commuted to work from rural areas. Commuting not only enabled the workers to live more cheaply than in the city but it also enabled them to help with farming or housekeeping on weekends. In addition, the maintenance of a rural residence provided a cushion during the economic recession when workers were laid-off. In the rural areas they could share in farm work with their relatives and thus avoid becoming a serious social problem for the urban areas (Wu, 1976).

The special role played by young women in the growing textile and electronic industries is described in case studies by Arrigo (1980) and Diamond (1979). Although many factories had initially built dormitories for single women, many women preferred to live with their parents and commute to work. Arrigo (1980) cites one factory in Taoyuan which sent out 58 buses daily to villages within an hour's drive to transport a work force of 4,000. These young women remit a large portion of their earnings to their parents. Diamond (1979) found that women in dormitories sent an average of 46 percent of their wages home and those living at home gave 70 to 80 percent to their parents.

Table 5.7 shows that time spent on travelling from the residence to the place of work per trip was relatively short with the majority of workers travelling less than 20 minutes and few travelling more than 40 minutes. With the exception of the Northern region which had fewer workers travelling more than 20 minutes, there was little variation in the distribution of travelling times by sex or region. These findings agree with our expectations. First, many firms and nearly all of the farms in Taiwan are owned and operated by family members. The residences of these family workers are usually shared with or adjacent to the business establishments. Second, in both urban and rural areas of Taiwan, residences and business sites are mixed and the transportation network is well developed so that those who work outside families could, without much difficulty, choose a combination of place of residence and means of transportation to minimize both the cost of living and the time to travel to work.

The previous finding that most of the geographical mobility of labor occurred at or near the time of entry into employment supports the above reasoning, for new entrants into the labor force are likely to be young,

TABLE 5.7

PERCENTAGE DISTRIBUTION OF TRAVELLING TIME EACH WAY BY REGION OF WORK, 1972

Region of Work, 1972	Less than 20 Minutes	20 to 39 Minutes	40 to 59 Minutes	60 Minutes and More	Total
		Male			
Total	67.6	23.8	4.9	3.7	100.0
Taipei City	63.0	26.5	6.5	4.0	100.0
Other Cities	65.2	25.7	6.7	2.4	100.0
North	75.3	18.2	3.6	2.9	100.0
Central	68.1	26.0	4.9	1.0	100.0
South	68.0	22.6	3.4	6.0	100.0
East	60.3	30.9	6.5	2.7	100.0
		Female			
Total	68.5	24.1	4.9	2.5	100.0
Taipei City	62.3	27.6	7.7	2.4	100.0
Other Cities	67.4	24.5	6.3	1.8	100.0
North	81.9	13.6	2.3	2.2	100.0
Central	62.7	31.1	5.5	0.7	100.0
South	69.8	24.8	3.7	1.7	100.0
East	60.4	33.8	4.8	1.0	100.0

unmarried persons and they can easily find a place to live nearby where they work.

The relationship between the means of transportation and place of work also supports this reasoning (see Table 5.8). also supports this reasoning. It indicates that the majority of people who worked outside large cities, had a residence near enough so that they could walk to work or go by bicycle or motorcycle. Females were more likely to walk; whereas, males were almost equally likely to walk or go by bicycle or motorcycle. Larger proportions of those who worked in cities, however, could live a little further from their place of work by taking a public bus or train. Government offices and government owned enterprises, mostly located in cities, generally provide transportation for their workers. Many large industries in rural areas also provide transportation. This phenomenon is reflected in a large proportion of workers using private buses and other means of transportation.

The Relation Between Industrial Mobility and Geographical Mobility

Further evidence in support of our argument that most of the changes from agricultural to nonagricultural employment did not involve moves to the large cities can be obtained from tabulations or geographical mobility for those persons who changed jobs or entered employment. These results, shown in Table 5.9, indicate that only 14 percent of those changing jobs or entering employment moved across township boundaries and only 3 percent moved to one of the large cities.[6] The figures were only slightly higher for those who moved from the primary sector to the secondary or tertiary sectors. Only 5 percent of these changes involved migration to one of the large cities.

Geographical mobility is nevertheless closely tied to job changes. The mobility rate for those changing jobs or entering employment was about three times that of persons who stayed at the same job. There was considerable variation by type of job change. In general, the geographical mobility rates were highest for people entering the tertiary sector and lowest for those entering the primary sector. These results indicate that job changes are an important source of variation in migration rates even though only a small proportion of those changing jobs move.

Job mobility shared the same relationships to age and sex as geographical mobility. Fifty-seven percent of those who changed jobs were under 35 in 1972 compared to 35 percent of those who did not change jobs.

----

[6] These rates would be somewhat higher if we added those who had moved without changing their legal residence, but this adjustment would probably not have much effect on the composition of the migration streams or the relative migration rates by type of employment.

TABLE 5.8

PERCENTAGE DISTRIBUTION OF MEANS OF TRANSPORTATION BY REGION OF WORK, 1972

Region of Work, 1972	Walk	Bicycle or Motorcycle	Public Bus or Train	Private Bus and Others	Total
		Male			
Total	44.1	40.2	10.0	5.7	100.0
Taipei City	19.5	37.6	32.4	10.5	100.0
Other Cities	27.6	53.5	12.1	6.8	100.0
North	59.7	26.8	7.7	5.8	100.0
Central	54.6	40.6	2.5	2.3	100.0
South	43.9	48.2	1.6	6.3	100.0
East	44.0	52.1	2.1	1.8	100.0
		Female			
Total	56.1	23.5	14.3	6.1	100.0
Taipei City	25.2	5.3	57.7	11.8	100.0
Other Cities	41.2	29.1	21.7	8.0	100.0
North	60.4	18.3	12.4	8.9	100.0
Central	69.2	26.2	3.0	1.6	100.0
South	57.9	30.9	4.6	6.6	100.0
East	62.2	32.0	2.6	3.2	100.0

TABLE 5.9

PERCENTAGE CHANGING RESIDENCE BY TYPE OF JOB CHANGE (thousands of persons)

1967 Industrial Sector	1972 Industrial	Number	Percent Moving to Different Township of City District	Percent Moving to Large City from Outside City
Not Employed	Primary	267	3.4	0.4
	Secondary	650	9.6	2.2
	Tertiary	515	15.6	4.2
Primary	Primary	29	27.6	1.4
	Secondary	89	17.7	5.4
	Tertiary	62	19.8	4.4
Secondary	Primary	12	8.2	0
	Secondary	122	19.6	5.1
	Tertiary	57	26.0	4.2
Tertiary	Primary	13	28.1	4.7
	Secondary	58	16.4	4.4
	Tertiary	182	31.1	4.8
Total Changing Jobs or Entering Employment		2055	14.4	3.2
Total Who Kept the Same Job		2915	5.1	0.7
Total 1972 Labor Force		4970	9.0	1.7

Four out of five of those changing jobs were males. Although more males were employed at both times than females, the rate of job change for those employed at both times was higher for males than for females.

A further factor contributing to high geographical mobility rates at young ages is the age distribution of new entries into the labor force. More than half of the entries to employment in the five year period were in the 15-19 age group in 1972. The median age in 1972 of those who had entered employment was about 18 for males and 19 for females. Since entry could have occurred at any time during the preceding five years, the estimated median age at entry was about 15 or 16 for males and 16 or 17 for females. The slightly higher median age of entry for females is due to the fact that more females than males leave and reenter employment. Although females comprised one-third of the labor force in 1972, they accounted for more than one-half of the entries to, and exits from, employment. The median age of females who left employment was only 29 compared to a median age of 59 for males.

Commuting Patterns in 1980

According to the 1980 Census, 73.6 percent of the workers in Taipei City also lived in the city (see Table 5.10). While this is an increase of 10 percent over the proportion observed in the 1972 survey, one must be cautious in assuming that this is a real increase. The 1972 survey used place of legal registration whereas the census used usual place of residence so that migrants who had not changed their legal residence were treated differently in these two measures. Nevertheless, it is possible that the magnitude of commuting to the central city may have declined due to the rapid growth of satellite cities and the movement of industries into these cities. We showed earlier that most of the net migration from 1975-80 was to Taipei County and not to Taipei City itself.

The six satellite cities which surround Taipei accounted for most of the residences of workers who commuted to Taipei to work. The rest came mainly from Keelung City and other parts of Taipei county; only 2.5 percent commuted from more distant counties or cities.

While the satellite cities around Taipei housed some of the workers in the central city, they also provided significant employment opportunities for their residents and persons living elsewhere. Out of 542,000 workers living in these six cities in 1980, 323,000 or about 60 percent worked in one of these cities.[7] Another 72,000 workers commuted to work in one of these cities from somewhere else. For example, Sanchung and Hsinchuang Cities

---

[7] Most of these workers (281,700) lived and worked in the same satellite city, but 41,300 commuted from a residence in one city to a workplace in another city.

TABLE 5.10

DISTRIBUTION OF PLACE OF RESIDENCE OF PERSONS WORKING IN TAIPEI CITY

	Number (thousands)	Percent
Taipei City	694.8	73.6
6 surrounding cities	172.2	18.3
Remainder of Taipei County	35.9	3.8
Keelung City	17.4	1.8
Elsewhere	23.2	2.5
Total	943.5	100.0

Source: 1980 Census.

drew workers mainly from the neighboring rural township of Luchou, while Panchio City drew workers from the rural town of Tucheng. Some workers were also reverse commuters who lived in Taipei City.

Similar commuting patterns were observed around other major cities. Kaohsiung attracted more than 20 percent of its workers from nine surrounding cities and towns with the neighboring city of Fengshan sending 43 percent of its employed residents (Population Census Office, 1982, p. 169). Tainan had commuting of 20 percent or more from three adjacent towns and Taichung had commuting of 20 percent or more from two towns.

As in the case of Taipei, the other major cities had nearby satellites which attracted separate commuting streams. The 1980 census identifies Fengyuan City which is just north of Taichung and Pingtung City which is east of Kaohsiung as two such places.

CONCLUSIONS

The transformation from an agricultural society to an industrial society typically involves massive rural to urban migration, rapid urban growth, and accompanying dislocations. In Taiwan, this transformation has occurred with less rural to urban migration and fewer urban problems than in many other countries. During the peak period of transformation, from 1955 to 1975, the proportion of the labor force employed in agriculture declined by about 30 percent, while the proportion employed in manufacturing, construction and related activities increased by 20 percent. There was considerable migration to the larger cities during this period, but this was not as great as it could have been because a significant amount of the industrial transformation took place in rural areas or involved daily commuting from rural residences to urban work places.

Rural industrialization has been important throughout Taiwan's development. The Japanese established food processing plants in rural areas during the 1930s. The basis for further rural industrialization was established by the land reform and rural development programs of the 1950s which gave small farmers a greater incentive to remain on their farms. As these farm households grew, they provided an increasing supply of labor which was not needed on the farm. In the 1960s and 1970s, specific government policies encouraged labor-intensive firms to locate in rural areas to take advantage of this labor supply and by the late 1970s over onethird of all factories were in rural areas.

The development of good roads linking rural and urban areas and the rapid increase in buses and other forms of public transportation facilitated commuting between rural and urban areas. By commuting, the workers were able to remain in their farm households and save the costs of housing in the cities. Studies of farm households showed incresed use of commuting during this period. Tabulations of place of residence by place of work from the 1980 census showed extensive commuting patterns in the Northern region of Taiwan involving Taipei City, satellite cities, and surrounding rural areas.

Finally, a survey of labor mobility in the 1970s showed that where there was migration from rural to urban areas that it typically involved young persons who were entering the labor force and that relatively few of the migrants were older farmers who were switching to nonfarm work. Most of these young persons had the basic education needed for work in the city and, as we shall see in later chapters, they had little difficulty in adapting to life in the city.

6

The Determinants of Migration to the City

In previous chapters, we have seen that the economic development of Taiwan has been accompanied by a movement of people from rural to urban areas. This movement corresponds to the decline in the proportion of the labor force employed in agriculture and to the large increase in the proportion employed in industry. Yet we have also seen that industrialization in Taiwan has been decentralized with a lot of the new manufacturing plants being located in rural areas. Furthermore, the excellent rail and bus lines make it possible for many workers to remain living in rural areas while commuting to work in urban areas.

In most cases the decision to move to city is one of choice and not necessity. There were no acts such as the enclosure acts in England which forced people who had been farmers to suddenly give up their occupations and search for new ones. Quite to the contrary, the land reforms of the 1950s made it easier for small farmers to remain in agriculture and to increase their income as well. These reforms also made a small amount of land available for landless groups such as retired military, but the limits of Taiwan's supply of arable land were soon reached. Between 1955 and 1975, the number of people employed in agriculture remained almost constant, while the labor force grew at rates of 2 to 4 percent.

The new entrants to the labor force had several opportunities open to them in most townships. They could seek employment as wage laborers in agriculture, they could seek nonfarm employment within their town, they could commute to work in a nearby city or town, or they could migrate from the town. Only in the most remote towns and those few towns which had not been penetrated by industrialization were these opportunities more restricted, and migration more a necessity and less a choice.

Given the range of alternatives, why did some people choose to migrate while others chose to remain in the town? In particular, given the geographic diffusion of development in Taiwan, why did some migrants chose to migrate to Taipei, the capital and largest city rather than move to a smaller city.

This chapter begins with a brief review of theories of migration followed by a discussion of the characteristics of the migrants and a comparison of the migrants with those who chose to remain in the rural areas. The data for these comparisons came from a 1973 survey of migration to Taipei which was conducted by the authors. This survey, which focuses on migration between 1908 and 1973 covers the latter part of the period of labor-intensive economic growth.

REVIEW OF PREVIOUS THEORIES

The migration theories which have most influenced policy makers are the cost-benefit theories proposed by economists. Sjaastad (1962) suggested that migration can be treated as an investment from which the migrant expects to receive returns sufficient enough to offset the costs of moving. The benefits from moving are expressed in terms of the difference in incomes between the place of destination and the place of origin for each future time period. These differences are then discounted to obtain the present value and summed over the total number of future time periods for which returns are expected. If the sum is greater than the cost of moving, the prediction is that the person will move.

While the theory is developed in terms of the individual decision-maker, most tests are conducted with aggregate level data. These tests have usually found that migration is related to the wage rate at the place of destination and sometimes to the rate of unemployment at the place of destination but rarely to the income or unemployment at the place of origin (Lowry, 1966; Beals, Levy and Moses, 1967; Sahota, 1968; Greenwood and Sweetland, 1972; Greenwood, 1975, 1985).

Todaro (1969) refined the cost-benefit model by adding a term for the probability of the migrant obtaining employment at the destination. With this model, Todaro was able to argue that migrants would still move to cities with high unemployment rates if there was a large enough gap between urban, modern sector wages and rural wages. What mattered was the expected income at the destination which was a product of the modern sector wage and the probability of finding modern sector employment.

There have been very few tests of the cost-benefit or Todaro models at the individual level. Speare (1969, 1971a) attempted to test these models with a sample of recent migrants to Taichung, Taiwan and a comparison group of nonmigrants living in the area of origin. He found that only a very small percentage of the migrants had heard about specific jobs in the city that paid well or could estimate their chances of getting a well-paying job. Instead, he found they had general expectations of increasing their income. Migrants were much more likely to have expected an increase in income at the time they moved than nonmigrants. These expectations were confirmed by increases in real income following the move. Migrants also paid significantly lower moving costs than those anticipated by nonmigrants. All of these results support the basic cost-benefit model, but not the more complex model developed by Todaro.

Speare's study also found that nonmonetary factors, particularly the location of relatives and the receipt of job information from friends or relatives, were also important determinants of mobility. A majority of the nonmigrants had not considered moving, and many of them could have increased their income by moving if one assumes that they would have been as successful as the average migrants. Shaw (1974) also found that a significant proportion of the population made no calculation of the costs and benefits from moving.

Temple (1975), in an analysis of the factors accounting for migration to Jakarta, concluded that most migrants were not responding to wage differentials. He stated, "the majority leave their village because they cannot find employment or because employment opportunities are uncertain. Once the decision to migrate has been made, migrants choose a destination where personal connections provide the most reliable access to activities yielding real income" (Temple, 1975: 81).

The main advantage of the cost-benefit model of migration is that it enables migration to be treated within the same general economic framework used for studying the labor market. Cost-benefit models provide a partial explanation for the observed relationship between migration and age and education (Schwartz, 1976). The Todaro refinement of the model also provides an explanation for the apparent contradiction between continued migration and urban unemployment.

However, the cost-benefit models do not adequately explain other relationships. The effect of distance on migration is usually greater than can be explained by the cost of moving. The models provide no clues for why migration rates should decline sharply with duration of residence. If the cost of moving were significant, we would expect to find migrants waiting a substantial time between moves. Where data are available, return migration appears to constitute a substantial proportion of all moves, yet the only explanation provided by economic models is that return movers are those who failed to achieve the level of income which they expected from moving.

The failure of cost-benefit models to explain these findings is probably related to the behavioral assumptions of the model. Studies of individual behavior have shown that migrants do not make precise calculations of costs and benefits. If they make any calculations, they are very crude ones. Monetary costs of moving are often insignificant compared to income gains in the first month alone. The probability of employment is too abstract a concept to enter into any calculations. Most migrants appear to find jobs soon after arrival and there is relatively little change from low-paying "informal sector" jobs to high-paying "modern sector" jobs in contradiction to the predictions of the Todaro model. Cost-benefit models also do not take into account the fact that most nonmigrants never consider moving or that friends and relatives are very important in determining both who moves and where they choose to move.

An alternative theoretical approach which focuses more on mobility behavior and particularly on the question of why some people never consider moving is found in the stress-threshold model of migration. This approach has its origins in the work of Simon (1957). Simon, in studying the behavior of business administrators, observed that decision-makers were limited in their capacity to formulate and solve problems and to acquire and retain information. To cope with these problems a decision-maker constructs a simplified model of the situation and acts rationally with respect to that model. Simon suggested that in this simplified model only a subset of the alternatives are perceived and payoffs are evaluated only as satisfactory or unsatisfactory. No action is taken if the current state is judged to be satisfactory. If it is unsatisfactory, a search is made for outcomes that are satisfactory and the search is terminated when a satisfactory alternative is found (Simon, 1957: 198-201).

This type of behavioral theory was first used for a study of residential mobility by Rossi (1955). Rossi saw the decision to move as depending on mobility potential, which was a function of household characteristics, and complaints with the current residence. Wolpert (1965) developed these concepts further. According to Wolpert, an individual will assign a "place utility" to the current place of residence which represents the social, economic, and other costs and benefits derived from that location. The individual evaluates this place utility relative to a threshold which is a "function of his experience or attainments at a particular place and the attainments of his peers" (Wolpert, 1965:162). Alternative locations are assigned utilities based on anticipated costs and benefits which may be imperfectly perceived. The range of alternatives is limited by one's awareness of opportunities elsewhere.

Brown and Moore (1970) have attempted to make this model more explicit by dividing the mobility decision into three stages. The first stage is the decision to seek alternatives which does not necessarily commit one to moving. This is followed by the search for alternatives, and the final evaluation which may result in the decision to relocate or to abandon the search and adjust to one's current location. They describe the search pattern as highly dependent on the information possessed by the individual and that available from the local environment. Since the information about a given opportunity is likely to decrease with distance from it, the final decision is likely to favor short distance movement over longer distance movement even where the costs of moving short versus long distances do not vary much. They also point out that the results of the first stage of the search process may feed back into the evaluation of one's current place utility and change one's satisfaction relative to the threshold for considering moving.

The behavioral approach was applied to residential mobility in the United States by Speare, Goldstein and Frey (1975). They substituted the notion of "residential satisfaction" for Wolpert's notion of "place utility." In a panel study of Rhode Island residents, they found that residential satisfaction acted as an intervening variable between individual and residential characteristics and residential mobility. Characteristics such as age, duration of residence, room crowding, education, income and city/suburb location were shown to have little or no direct effect on mobility once residential satisfaction was controlled. The only characteristic which consistently showed a direct effect on mobility was home ownership. The negative effect of home ownership was explained in terms of the higher costs of moving for homeowners in comparison to renters.

Bach and Smith (1977) have successfully used this model to explain intercounty migration in the United States. They expanded the model somewhat to include additional measures of satisfaction. They found that the index of community satisfaction, which was constructed from questions about the respondents' image of the community, was one of the strongest predictors of migration. In a study of interstate migration, Speare, Kobrin and Kingkade (1982) found that satisfaction with community and job did not have much effect on migration, but social and economic bonds at the place of origin did inhibit migration.

A THEORY OF INDIVIDUAL MOBILITY

The theory of migration presented here combines elements of the cost-benefit and stress-threshold models in a way that can be operationalized for the study of individual mobility. Following Brown and Moore (1970), migration is viewed as the result of an ongoing decision-making process for which three stages can be distinguished: (1) the development of a desire to consider moving, (2) the selection of an alternative location, and (3) the decision to move or to stay. However, these stages are not distinguished in the operational model because of the problems associated with measuring the outcome of stages (1) and (2) independently of the outcome of stage (3). Mobility decisions often take place within a short period of time and it is almost impossible to observe each of these stages separately. Even if such observation were possible, it is difficult to formulate simple mathematical expressions relating these stages (see Speare, Goldstein and Frey, 1975:183-205).

In order to simplify the model, it is assumed that the decision to move or stay is made by a single decision-maker who takes the varied interests of other household members into account. While many decisions may actually be the result of a complicated interplay of interests within a household and may sometimes involve the division of a household, these aspects of decision-making are not well enough understood to be represented in the model at this time.

The decision to move is hypothesized to depend on four types of factors which are central to the cost-benefit and stress-threshold models. These are (1) social and economic bonds at the place of origin, (2) residential satisfaction and job satisfaction at the place of origin, (3) awareness of opportunities elsewhere, and (4) expected costs and benefits from moving. Each of these factors is described below.

Social and Economic Bonds

Individuals are tied to their place of origin by bonds to other individuals, attachments (both economic and psychological) to a particular piece of land or housing unit, attachment to a job, and other local bonds. The more people there are in a household, the more bonds there are likely to be. Households with strong bonds to a place of origin may not consider moving even though they might be better off somewhere else were they to calculate the costs and benefits. If the family farm cannot provide adequate employment for all family members, these members can seek other employment in the villages or commute to the city for work without having to break their bonds to the village.

Satisfaction

In the three stage model it was postulated that in most cases of voluntary mobility the initiation of the decision-making process results from a person's increasing dissatisfaction beyond his threshold or tolerance level. This concept is essentially the same as the stress-threshold concept used by Wolpert (1965). The term *dissatisfaction* is preferred to the term *stress* because dissatisfaction avoids the connotation of mental tension. Two major sources of dissatisfaction are aspects of one's place of residence and aspects of one's job. For example, a person may consider moving because of dissatisfaction with the size of one's house, with the neighborhood, or with the quality of nearby schools or with aspects of one's job such as income, working conditions, or freedom to work independently. While the three stage model postulated that dissatisfaction was a necessary condition for mobility, it is difficult to measure satisfaction with sufficient precision and at the right point in time to be able to assign it such a central role in the model. Because of these measurement problems, we shall merely treat satisfaction as one of the factors affecting mobility and allow for the possibility that people who are observed to be satisfied where they are may move because of other factors.

Awareness of Opportunities

Brown and Moore (1970) placed considerable emphasis on the role of knowledge of alternative locations in the mobility decision-making process. They defined the concept "Awareness Space" as the space which included locations that the potential migrant has knowledge about through previous residence or other direct contact, through friends and relatives, through the mass media or through contact with specialized agencies such as employment services. In most cases, awareness space is limited to a few locations. If a person has lived in the same place since birth, traveled little, and has similar friends and relatives, that person's awareness space may contain no real alternatives. On the other hand, a person who has lived in several other places or who has friends and relatives in other places can consider these places as alternatives. Migration studies have frequently shown a strong tendency for migrants both to return to places of previous residence and to move to places where they have friends and relatives. The prediction then is that the probability of moving will increase with the size of one's awareness space if all other factors are held constant.

Costs and Benefits of Moving

The costs and benefits of moving have been discussed at length by economists (DaVanzo, 1976; Greenwood, 1975; Sjaastad, 1962; Todaro, 1976). Although these always include the increase in income of the job at destination over that at origin and the out-of-pocket costs of moving, the list can be expanded to include other increased benefits such as better health care facilities, better schools and other facilities. If any of these are inferior at the place of destination compared to the origin, they count as negative benefits. Economists sometimes also refer to "psychic costs" arising from the separation from friends and relatives and other factors. These need not be included under costs and benefits here, however, since we have separate factors for social and economic bonds and satisfaction.

Individual Characteristics and Migration

The basic hypothesis of the model is that mobility is a function of the four factors described above. The corollary is that individual, residence and job characteristics do not have an independent effect on migration once these factors are controlled but affect migration indirectly through their effect on these four basic factors (see Figure 6.1). For example, migration has observed to have been a strong relationship to education in Taiwan and most other developing countries. This relationship could act through any four of the major factors. The most frequently discussed mechanism is that persons with higher education can expect to find greater differences between the income offered by urban versus rural jobs than those with lower education. However, it is equally likely that persons with higher education will have larger awareness spaces. Lansing and Mueller (1967:214) observed in the U.S. that the number of alternatives considered and the number of sources of information used increased with education. Increased education may also orient a person towards an urban way of life and make them less satisfied with rural life. Finally, higher education may increase one's contacts with people outside the village and thereby weaken one's dependence on social bonds in the village.

Age has opposing effects on the major determinants of mobility. While awareness space tends to increase with age, social bonds also tend to

FIGURE 6.1

CAUSAL MODEL OF DETERMINANTS OF MIGRATION

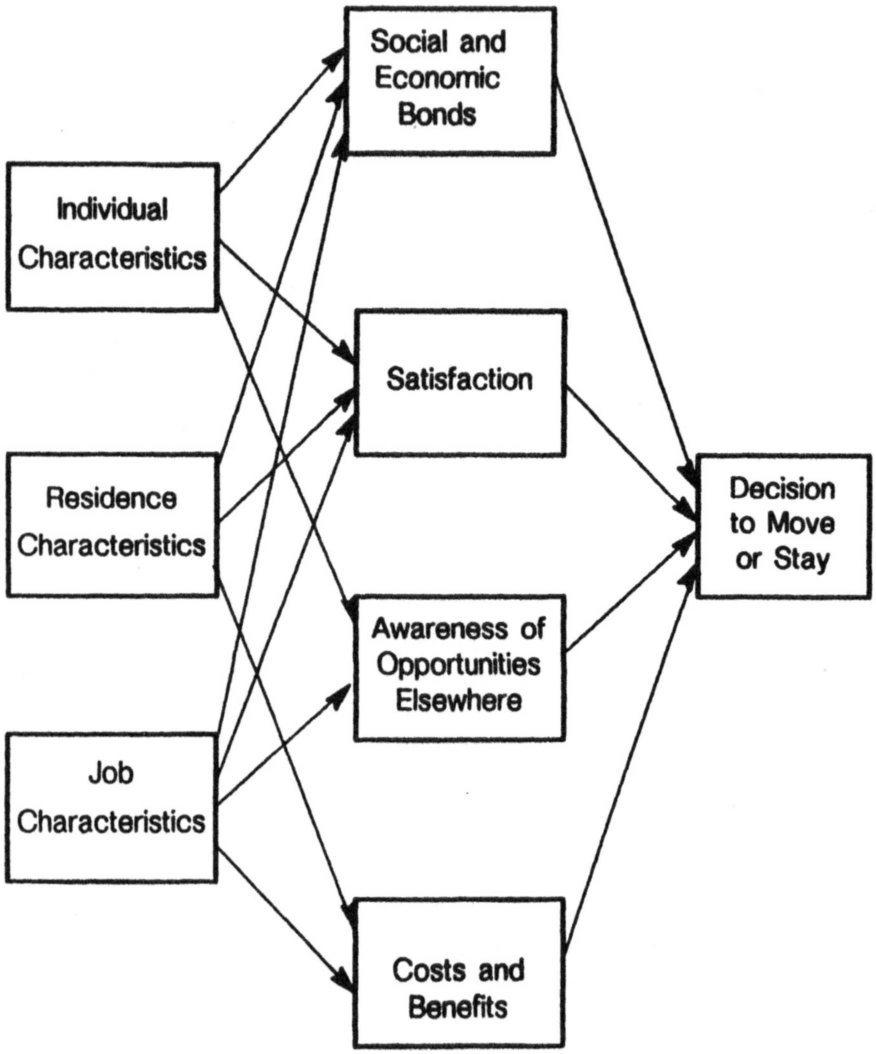

increase. The benefits from moving probably decrease with age as one acquires locally specific human capital which cannot be transferred to another location. Clearly the number of years which one can receive the benefits at the place of destination decreases with age. Costs of moving an entire household increase with the age of the head as family size increases and furniture and other possessions are acquired. Overall, there are more factors acting to decrease mobility with increasing age than factors acting to increase mobility.

Characteristics of housing and land also affect economic bonds and residential satisfaction. Property owners tend to be more satisfied than renters because of the pride and social prestige in owning their own home or land. In addition, the perceived difficulties in selling a house or land tend to make opportunities elsewhere seem less attractive. Tenant farmers who have acquired certain rights in the land over time may also be reluctant to move.

Jobs may or may not tie people to particular locations. Farmers with land and employees of enterprises with one fixed location have economic bonds which they must sever if they permanently leave the location of these jobs. However, many farm laborers do not have regular employment and many members of the nonfarm labor force are self-employed. These people still have some economic bonds to the extent that they have employers or customers who give them preference. These bonds do not restrict short distance residential mobility, but they do enter into the determination of satisfaction with the area of residence and thereby affect inter-area mobility.

Social and Economic Change and Migration

If everyone made the right choice of residence and if all the factors associated with that residence stayed the same, our discussion up to this point would lead to the conclusion that there should be no mobility. In fact, increasing duration of residence and increasing age would serve to increase satisfaction which would act to make mobility even less likely as time increased.

However, most societies today are in the process of change. New job opportunities are constantly being created while other jobs are eliminated. Population growth forces children in some families to seek employment outside the family farm. Improvements in transportation make it possible for people to travel more easily and to learn of opportunities elsewhere. All these factors, and many others, can affect an individual's evaluation of current place utility and lead to the consideration of migration.

Although the model represented in Figure 6.1 may appear to be a static one, all of the above changes can operate through the model. For example, a deterioration in land quality over time should lower the residential satisfaction of the residents and lead some of them to consider moving. A model that uses levels as the independent variable is better than one that uses changes as the independent variable for two reasons. First, residential satisfaction is likely to be more a function of the level of the independent variables than a change in these variables. Second, households are often slow to adjust to changes, and it would be difficult to define the appropriate period of time over which to measure changes in the dependent variables.

Another type of change occurs as people progress through the life cycle. At the points of transition between life cycle stages some of the old bonds are weakened and new needs are created which force a person to take some action. The most important example is the transition from youth to adulthood when a person ceases to be entirely dependent on his or her parents and becomes an independent decision maker. This usually occurs about the time when one enters the labor market on a relatively permanent basis, but may be delayed if the parents can provide employment on the family farm or in the family business. Those who enter the labor market without prescribed jobs frequently consider mobility at that time. This can be explained in terms of weakened bonds to the place of residence and decreased satisfaction resulting from a new unfulfilled need - the need for employment. If local employment can be found which satisfies one's expectations, then the bonds are strengthened and residential satisfaction quickly increases. If satisfactory employment cannot be found, the alternative of mobility is actively considered.

In many rural areas of developing countries there are no schools beyond elementary school within the range of daily commuting. If further education is desired, a person must move to a city or town. When the education is completed (or otherwise terminated), that person has acquired some ties to the urban area. The person then faces the choice of moving back to the village or staying in the city. The theory predicts that if the person is satisfied with other aspects of city life, that person will first search for a job, a spouse, or other means of support in the city before considering returning to the village.

THE MIGRATION SURVEY

One of the major objectives of the 1973 Taiwan Migration Survey was to obtain measures of the variables in the model of individual migration to enable that model to be tested. The survey, which is described in greater detail in Appendix A, was designed to provide a comparison of these measures between recent migrants to Taipei City and residents in a sample of urban and rural areas in the rest of Taiwan that represented the areas of origin of the migrants. The residents outside Taipei can be considered nonmigrants in the sense that they had not moved to Taipei within the five year period used to define recent migrants. However, they may have made other moves during this period and could have lived in Taipei during some earlier period. Both samples were restricted to persons aged 20-39 at the time of the survey.

In addition to measuring variables in the model of individual migration, the survey also obtained information about the reasons for moving. Most of the male migrants said that they came to Taipei to find a job or because of a change in location of an existing job.[1] The proportion moving for job related reasons was 70 percent among male migrants, but only 30 percent among female migrants. Nearly one-half of the female migrants said they moved for marriage or to follow their husband or parent. Ten percent of the males and nine percent of the females moved to the city to further their education. A small number of both males and females gave housing or other reasons for moving.

Recent migrants differed in age from residents at the places of origin. Within the 20 year age range of the samples, migrants tended to be younger than residents (see Appendix Table A.2). Among recent migrants, 41.5 percent were under age 25 compared to 27.0 percent of the residents of urban towns and 20.6 percent of the residents of rural towns. The age differences are even greater if one uses the age at move in the comparison (see Appendix Table A.3). Since migrants could have moved up to five years before the survey, nearly 58 percent were under age 25 at the time of the move.

Migration of persons aged 20 to 39 does not appear to vary greatly by sex. The sex ratio for migrants from urban areas is identical to the sex ratio for residents of these areas (.73) and the sex ratio for migrants from rural areas (.92) is only slightly greater than that for rural residents (.88). All sex ratios indicate a deficit of males which can be attributed to (1) the absence of males who are serving in the military and (2) the higher response rate for females in the survey.

There are substantial differences between migrants and residents in education (see Table 6.1). Education through at least primary school is the norm for people in this age range in all areas; over ninety percent of the migrants had completed primary school compared to eighty-eight percent of the residents outside Taipei. The largest differences are for males from urban areas where two-thirds of the migrants had completed senior high school in contrast to only about one-fifth of the residents who remained in urban areas. The differences between male migrants from rural areas and residents of these areas are also great. Female migrants were neither as highly educated as male migrants nor as different from those who remained behind.

In the following section we shall try to show that age and education selectivity of migration is related to differences in social and economic bonds, satisfactions, information, and perceived costs and benefits of migration among people of different ages and levels of education.

---

[1] These included six migrants who were serving in the military, but not located in barracks.

TABLE 6.1

EDUCATION DISTRIBUTION (highest level completed) of MIGRANTS AND RESIDENTS IN AREAS OF ORIGIN BY SEX AND TYPE OF AREA

	Migrants		Residents	
	From Urban Areas	From Rural Areas	In Urban Areas	In Rural Areas
Males				
Not Primary Grad.	4.3	8.3	6.2	21.6
Primary Grad.	19.6	33.3	55.8	63.8
Junior Grad.	8.7	16.7	18.6	8.6
Senior Grad.	37.0	33.3	14.2	4.3
College Grad.	30.4	8.3	5.3	1.7
Total: Percent	100.0	100.0	100.0	100.0
Number	46	24	113	116
Females				
Not Primary Grad.	13.3	11.5	18.8	31.1
Primary Grad.	40.0	42.3	51.9	50.8
Junior Grad.	15.0	7.7	9.7	8.3
Senior Grad.	20.0	26.9	13.0	9.1
College Grad.	11.7	11.5	6.5	0.8
Total: Percent	100.0	100.0	100.0	100.0
Number	60	26	154	132

The relatively small numbers of recent migrants in this survey make it difficult to divide them by type of place of origin and sex as was done in the first two tables. In the remainder of this chapter, migrants from all areas of origin will be combined together and both sexes will be shown together except where sex makes a significant difference in the results. To provide a proper comparison with the sample of residents, the residents will be weighted so that they have the same distribution of type of place as the distribution of type of place of origin of the migrants. The weighting will be done in such a way that the total number of cases remains the same. In addition, multiple classification analysis will be used to adjust the results for differences in age, sex and education.

TEST OF THE THEORY OF INDIVIDUAL MOBILITY

The first step of the analysis will be to determine which of the theoretically important factors has an effect on mobility for this sample. The importance of each factor will be assessed by examining the difference between migrants to Taipei and residents at the place of origin. Where there is a difference we can infer that the variable may be a determinant of mobility. Since all of our data was collected after the fact, we cannot prove that a particular variable was a cause of mobility. Nevertheless, the results of these tests will indicate which parts of the theory are plausible, which ones cannot be supported, and which ones need to be better operationalized before any test can be made. This analysis is divided into four sections which deal with, in order, bonds to the previous place of residence, satisfaction, awareness space and the expected costs and benefits from moving.

Bonds to the Previous Place of Residence

The theory predicts that social, economic, and other bonds act to hold a person to a particular location. The stronger the bonds the lower the chance that a person will move, or even consider moving. In this section we shall compare the social, economic, and housing bonds of migrants to Taipei prior to moving with those of residents outside Taipei.

The first bond to be considered is marriage. When a single person becomes married the number of bonds are increased. If the spouse is from the same township, as is most common in Taiwan, the bonds holding the married couple to the town are roughly double those holding the single person. While some married persons do migrate without their spouses, such separations are far less common in Taiwan than in India and parts of Africa. In an earlier study of migrants to Taichung, Speare (1969:112) found that most of the married male migrants moved with their spouse and that by the time of interview, which was one to two years after the move, 96 percent of the wives were living with their husbands in the city.

The effect of marriage in retaining people in the smaller urban and rural areas is seen in the distribution of migrants versus residents by marital status (Table 6.2). About two-thirds of the migrants were married at the time of moving compared to about four-fifths of the nonmigrants remaining in the areas of origin. The higher proportion single among migrants is related to the fact that migrants are younger than residents, but the relationship becomes reversed when age, sex, education and rural/urban origins are controlled. When examined separately by sex, the unadjusted relationship between migration and marital status is stronger for males than females, although the adjusted relationship is not statistically significant. Among females, the unadjusted difference is small but the difference became larger when age, education and type of place of

TABLE 6.2

MEASURES OF SOCIAL BONDS FOR MIGRANTS AND RESIDENTS IN THE AREAS OF ORIGIN

	Unadjusted		Adjusted	
	Migrants	Residents	Migrants	Residents
Percent Currently Married	67	80	85	75***
Males	47	74	73	67
Females	83	84	95	81***
Percent living in same household as parents	66	61	56	64
Percentage of cases where respondent or spouse could decide how to spend family income	28	60	42	56***
Percentage owning land	32	41	36	40
Percentage owning 2+ hectares	1	7	2	7***
Percent Employed	50	59	51	59*
Males	61	96	76	91
Females	39	29	37	29
Percentage of those employed who worked on farms	13	24	22	22
Percentage of employed who were self-employed	6	24	12	22***
Percentage who owned homes	54	76	56	75
If respondent or spouse head	31	71	39	70***
If parent or relative head	63	81	65	80***

Adjusted measures are adjusted for age, sex, and education using multiple classification analysis.

*significant at p=.10 level, **significant at p=.05 level, ***significant at p=.01 level.

origin are controlled, with migrants being more likely to be married than nonmigrants. This may be because many married women move to accompany their husbands.

The strongest social bonds outside the nuclear family are to the parents, usually those of the husband. In fact, many people in Taiwan still adhere to the old Chinese norm that a man should feel closer to his parents than to his wife (see Speare, 1974b).

Prior to moving, most migrants lived in households containing their parents. This is not surprising since many were unmarried at the time of moving and since it is the custom for married sons to live with their parents for at least a few years following marriage. Approximately two out of three migrants lived with their parents before moving (see Table 6.2). This information was obtained from a question on household composition for each place in the migration history. When adjusted for age, sex and education the proportion of migrants who lived with parents before moving is slightly less than the proportion of residents but the difference is not statistically significant. This means that the absence of a close bond to parents was not an important factor in determining migration.

It is likely that many young adults are actually torn between the desire to remain with their parents and the desire to form their own household and gain control over household expenses. These contradictory goals often lead to conflict with Chinese families. The conflict can be resolved in two ways. Either the parents can give over the control of household expenses, or at least the portion of expenses not used directly for themselves, to their children or the children can move out and establish a separate household. If the latter solution is chosen, the need to move opens up the consideration of moving longer distances. In fact, in some instances it may be easier to explain the situation to one's parents if one moves to the city rather than across the street.

The relationship between the control of household income and migration is shown in Table 6.2. Migrants were far less likely to control household income before moving than residents in the areas of origin. Even when the results are standardized by age, significantly fewer migrants controlled household income before moving than residents in either cities or rural and small towns. After moving 53 percent of the migrants controlled their household finances, a substantial gain. These results suggest a qualification of the theory relating social bonds to mobility. It appears that social bonds which place a person in a subordinate position may encourage mobility.

A major economic bond to the area of origin is the ownership of farmland. Before moving, approximately 32 percent of migrants to Taipei lived in households that owned farmland, but this is only slightly less than the 41 percent of residents who owned farmland (see Table 6.2). These percentages become somewhat closer when adjusted for the control variables so that the differences between migrants and residents are insignificant. The finding that there is no significant difference between migrants before the move and residents in the area of origin in the ownership of farmland is in contradiction to that of the 1967 Taichung Survey where migrants were found to be less likely to own land before moving than residents in the area of origin (Speare, 1969). The implications are that the ownership of farmland per se is no longer an important barrier to migration in Taiwan. Most of the farm families owned less than one hectare of land. With rising urban wages, the returns from farming small plots of land were probably not as good as those from urban employment. In many cases there were other household members who continued to operate the farm after the migrant left and where there were none the farm could be sold or rented.

Only about one percent of the migrants came from farms with two or more hectares. This percentage was considerably lower than that for the residents (7 percent) and that difference was significant even after adjustment, indicating that the ownership of relatively large amounts of farmland does provide a barrier to migration. However, only a small proportion of rural and small farm residents are large landowners so this cannot be a major factor in explaining the immobility of the rest.

A second type of economic bond is employment. Because employment differs significantly between males and females, we have shown results separately for each sex. We have also included information obtained about spouses of married respondents. Male migrants were less likely to have been employed before moving than were male residents at the time of interview. Only 61 percent of the male migrants were working compared to 96 percent of the residents (see Table 6.2). Male migrants were more likely to be unemployed or in school before moving than were residents.

Female migrants had a higher proportion employed before moving than female residents but also had higher proportions unemployed and in school than the residents. While the relationship between employment and migration for females appears to contradict that for males, this may be a consequence of the female migrants being younger and less likely to be married than female residents. Labor force participation of women in Taiwan is high between the time of leaving school and marriage, but drops dramatically after marriage. If marital status is controlled the single female migrants were less likely to have worked than the single residents which is consistent with the theory. The married female migrants have slightly higher rates of employment before moving than residents, but some of these may have become married only after their move.

The strength of the economic bond provided by employment varies with the type of employment. Among those employed, migrants were less likely to have been farmers before moving than the residents (see Table 6.2), although this difference disappears when age, sex education and type of place are controlled. Migrants were also less likely to have been self-employed than residents and this difference remains significant after adjustment. This indicates that the bonds to clients or customers may be stronger than those to an employer. Put another way, it may be easier to leave one employer and find another than it is to move one's own business to a new area and seek new clients or customers.

Male migrants were more likely than residents to have been government employees before moving than male residents and female migrants were about as likely as residents on the average. Because many government jobs, are national jobs it is relatively easy for government employees to move from one location to another. In fact, the rotation of civil servants has long been a common practice in China.

In addition to land and employment, the ownership of a house can provide an economic bond deterring people from moving. In the analysis of this question, it is important to distinguish whether the house was owned by the person in question or by the parents or other relatives. This is done in Table 6.2 by controlling for the relation to the head of household. In both cases migrants were less likely to live in homes which were owned by the household than residents. The difference was very large for migrants who had been the head of household or spouse of the head before moving. Only 31 percent of these owned their house compared to 71 percent for residents outside Taipei and most of this difference remains after adjusting for other variables. Home ownership appears to be an important bond which affects mobility.

Satisfaction at Previous Place

Two measures of satisfaction were obtained. The first of these dealt with job satisfaction and was obtained from a single question which asked migrants how satisfied they were with their job before they moved to Taipei. Residents were asked a similar question about their current job to enable a comparison to be made. Both groups were given a choice of five response categories ranging from "very satisfied" to "very dissatisfied." The results, which are shown in Table 6.3, fail to support the hypothesis that job dissatisfaction is one of the factors contributing to migration. For both males and females, the migrants reported slightly higher levels of satisfaction with their prior jobs than residents reported for their current jobs. However, these differences were too small to be statistically significant and were opposite to what was predicted. While it appears that job satisfaction does not have much affect on migration, our single question measure may be grossly inadequate or the response for migrants may have been biased by the recall period which could have been as long as five years for some of the recent migrants.

The second satisfaction index deals with the place of residence and is constructed from summing the responses to six questions which inquire about satisfaction with the size of the house, the physical condition of the house, the neighborhood, the distance from work, the rent or cost, and the quality of nearby schools. As with job satisfaction, migrants were asked about their place of residence prior to moving and residents were asked about their current place of residence. Each item was scored from one to five with one representing "very satisfied" and five representing "very dissatisfied" and the scores for the six items were summed to form the index.

The results which are shown in Table 6.3 indicate that migrants had levels of residential satisfaction which were similar to those of residents. The adjusted index value for migrants is slightly higher than that for residents indicating greater dissatisfaction, but this difference is not significant. The lack of results could be due to the interpretation of "satisfaction" in Chinese culture. Very few respondents indicated dissatisfaction and this could be due to their attempt to give polite answers. Also the gap of up to 5 years between the period referred to and the interview may have made it difficult for respondents to accurately recall their feelings of satisfaction. In a study of residential mobility by Speare, Goldstein and Frey (1975), where residential satisfaction was found to be related to mobility in the United States, the measure of residential satisfaction was obtained in an interview prior to mobility.

Awareness Space

The theory predicted that persons who were familiar with other locations would be more likely to move than those without such familiarity and that the likelihood of migration would increase with the number of locations with which persons were familiar. The size of each respondent's awareness space was measured using a series of questions similar to those employed by Brown and Associates (1972) in a study of migration in Ohio. The interviewer read a list of cities and for each city the respondent was asked "please tell me how familiar you are with it." The list included the five largest cities in Taiwan plus the four cities over 100,000 which were near Taipei City and might be alternatives for persons considering moving to Taipei. In addition, the respondents were asked to name any other cities with which they were familiar. The city of residence, if they were living in a city, was not included. For each city with which the respondent was familiar, questions were asked on the type of contact they had with the place and how attractive a place they thought that it was to live in.

Most respondents were familiar with three or four cities in addition to the place where they were living at the time of the interview. The awareness space was larger for migrants than for residents on the average (see Table 6.3). However it tended to be larger for persons in their twenties than those in their thirties and for those with higher levels of education. When these variables were controlled, there was no consistent difference between migrants and residents of areas outside Taipei. Since Taipei was not counted as part of the awareness space for migrants, the mean of 3.4 indicates that they were familiar with at least three other places on the average.

The degree of familiarity varied with the type of contact which they had had with places. For example, among migrants who were familiar with Taichung City, about 36 percent had lived there at some time in the past, 20 percent had relatives or friends living there and the remaining 44

TABLE 6.3

MEASURES OF SATISFACTION, INFORMATION, AND EXPECTED COSTS AND BENEFITS FOR MIGRANTS AND RESIDENTS IN THE AREAS OF ORIGIN

	Unadjusted		Adjusted	
	Migrants	Residents	Migrants	Residents
Average Job Satisfaction (1=high to 5=low)	2.74	2.99	2.86	2.96
Average Residential Satisfaction (6=high to 30=low)	14,8	14.8	15.0	14.7
Awareness Space Measures				
1. Number of familiar cities	3.4	2.6	2.9	2.8
2. Cities with personal contact	2.0	1.6	1.8	1.6
3. Attractive cities	.8	.6	.7	.7
Percent with Job Information from Friends and Relatives	47	19	45	19***
Percent with Information on Living Conditions	34	22	28	24
Expected Change in Income from a Move to Taipei (1=a lot more to 7=a lot less)	2.8	3.5	3.0	3.4**
Proportion Expecting Cost of Moving to Exceed NT$ 1000.	45	85	44	85***

Adjusted measures are adjusted for age, sex and education using multiple classification analysis.

**significant at p=.05, ***significant at p=.01.

percent had only visited the city briefly. In Table 6.3, we have shown the mean number of cities with which respondents had "personal contact," which we defined as respondents having lived there previously or having friends or relatives currently living there. Again, migrants had personal contact with more cities than did nonmigrants, excluding the city of current residence, but the difference was not significant when other variables were controlled.

Not all of the places with which the respondents were familiar were seen as attractive places. The average number of attractive places varied from .6 to .8 compared to a range of 2.6 to 3.4 for familiar places. There were no significant differences between migrants and residents of areas of origin.

The results of this investigation are inconclusive. The migrants are familiar with more places than the nonmigrants but do not differ in the number of cities which they view as attractive. However, we have biased the results somewhat by excluding Taipei, the place of destination. If we were able to assume that all migrants to Taipei had been familiar with the city then we could add one place to the count for migrants. If we could also assume that they thought of Taipei as an attractive place to live, we could also increase the count of attractive places by one. This would give the migrants significantly larger awareness spaces than the nonmigrants. There is also the possibility of an opposite bias arising from the assumption that the awareness space measured at the time of interview was the same as that at the time the decision to move was made, when, in fact, it might have been somewhat smaller. Finally, it is doubtful whether the size of the awareness space would be a decisive factor in determining mobility. Most residents outside Taipei were familiar with at least one city, other than the one in which they were living, which they regarded as an attractive place in which to live. If awareness space were decisive, they would have also moved.

If we move from the notion of general familiarity and attractiveness of a potential place of destination to the possession of knowledge about specific job opportunities at that location, the results become more conclusive. Both migrants and residents outside Taipei had received information about job opportunities from a variety of sources. About one-half of the migrants had received job information from friends or relatives. The only public source of information which was used was the newspaper and this was utilized by only 16.3 percent of the migrants. Where both public and private sources of information were mentioned, the migrants usually reported that the information received from private sources was more useful than that received from public sources. When asked to evaluate the accuracy of their information, 60 percent of the migrants thought it was very accurate, 33 percent thought it was mostly accurate, and only 7 percent thought it was inaccurate.

The proportion of migrants who had received information from friends or relatives was more than double that for residents in urban and rural areas even when age, sex education and type of place of origin were controlled (see Table 6.3). These results support the theory that access to information is an important factor in determining who moves.

Migrants were also more likely than residents to have recent information on living conditions in Taipei prior to their move. The majority of both migrants and nonmigrants had made trips to Taipei, although only a small proportion of these trips were for the purpose of looking for work.

Costs and Benefits of Moving

Most of the economic literature on migration has emphasized the monetary costs and benefits. Central to the economic approach is the concept that migrants are motivated to move by expected increases in income. Previous research by Speare (1969) showed that few migrants had precise estimates of their likely gains in income before they moved, although most expected some increase in income. However most of those who were in the labor force had a general expectation about income. We measured this on a seven point scale ranging from 1 for those who expected to make a lot more money in the city to 7 for those who expected to make a lot less. The average value for migrants was 2.8 which is somewhat above "a little more" compared to nonmigrants who had an average of 3.5 which was between "a little more" and "the same amount" (see Table 6.3). This difference got somewhat smaller, but remained significant, when the background variables were controlled. Many nonmigrants had not given this question any previous thought, which is reflected in the high proportion who replied "don't know."

Another important factor is the expected cost of moving. Table 6.3 compares the proportions who expected the cost of moving to exceed NT$ 1000. Residents on the average expected higher costs than the migrants experienced. Whereas only 45 percent of the migrants had paid more than NT$ 1000 (about U.S. $26), 85 percent of the residents outside Taipei thought that a move would cost more than NT$ 1000. All respondents were asked for the precise cost if they knew it or could estimate it but only 17 percent of the migrants and 9 percent of the residents could give precise cost figures and the accuracy and representativeness of these are doubtful. They ranged from no cost to about NT$ 50,000. The median for migrants was about NT$ 450, while that for residents was about NT$ 3500. While lacking precision, the data on costs of moving indicate that this is an important factor separating migrants from nonmigrants.

Multivariate Analysis

In the preceding analysis, we have examined the relationships between a large number of variables and migration and have found that several of these variables are significantly related to migration. We are left with two questions: (1) to what extent do these variables have independent effects on migration as opposed to simply representing different measures of the same effect; and (2) what is the combined effect of these variables on migration? These questions can be broken down into a series of more detailed questions about specific sets of variables and issues such as the relative effect of social and economic bonds versus intervening variables on migration.

In this section we shall employ logistic regression to attempt to answer these questions. Logistic regression is an appropriate technique when the dependent variable is dichotomous, as is the case with migration.

A total of 17 variables have been included in the multivariate analysis. These include four background variables, six measures of social and economic bonds and seven intervening variables representing satisfaction, awareness of opportunities and expectations of costs and benefits. Some of these variables are missing for particular cases and the missing data are not concentrated in a few cases, but spread throughout the sample; this presents a problem in carrying out the analysis. We have followed the procedure of deleting all cases with three or more missing items.[2] This has resulted in deleting 80 cases from the analysis. When variables with missing values were included in the analysis, the missing cases were allocated to the "no" category in the case of dichotomous variables such as job information or to the central category in the case of the satisfaction variables.

The results of logistic regression are presented in Table 6.4. The variables are shown in three groups: background characteristics, social and economic bonds and intervening variables. When entered in this order, the background variables explain 15 percent of the baseline Chi-square, the bonds explain an additional eight percent and the intervening variables explain another 11 percent for the total sample.[3] While these percentages depend on the order in which these blocks of variables are added, a reordering of the blocks so that each block is last does not change the fact that each block makes an independent contribution to the explanation of migration. When all the variables are entered, the residual Chi-square is not statistically significant which means that it is not necessary to add interaction terms in order to get an adequate fit to the data.

The theory predicts that background variables will not have a

---

[2] A exception was made for job satisfaction which had missing values for all persons who were not employed and for some who were employed. For these cases, the value of "indifferent" was assumed and job satisfaction was not included in the count of variables with missing values.

[3] The baseline Chi-square is obtained by entering only the constant term. While somewhat similar to R-square in multiple regression, this measure is not the same because it depends upon the variables in the final model (Goodman, 1972).

TABLE 6.4

LOGISTIC REGRESSION COEFFICIENTS FOR DETERMINANTS OF MIGRATION

	Model A	Model B	Model C
A. Background Characteristics[(1)]			
Sex (Male)	.148	.317**	.332**
Age at Move: 15-24	.771***	.546**	.553**
25-29	.321	.296	.250
30-34	-.377	-.226	-.189
35-39	-.715	-.616	-.614
Education: Less than 5 years	-.358***	-.511***	-.548***
Primary Grad	-.484	-.425	-.405
Junior High	-.018	.046	-.127
Senior High	.860	.890	1.080
Urban/Rural Origin (Urban)	.331**	.173	.245*
B. Social and Economic Bonds			
Marital Status at Move (Married)		.267*	.618***
Decision Maker (R or Spouse)		-.527***	-.572***
Homeowner (Yes)		-.612***	-.656***
Land Owned: None		.490**	.900**
Less than 2 hectares		.713	.811
2 or more hectares		-1.203	-1.711
Job at Origin (Yes)		-.105	-.152
Self-employed (Yes)		-.634***	-.673***
C. Intervening Variables			
Residential Satisfaction			
Satisfied			-.002
Indifferent			.341
Dissatisfied			-.339
Job Satisfaction:			
Satisfied			.172
Indifferent			.065
Dissatisfied			-.237
Awareness Space (3+ cities)			.084
Job Information (Yes)			.528***
Living Conditions Info. (Yes)			.139
Expected Increase in Income (Yes)			.469***
Expected Moving Costs (Yes)			-.811***
Constant	-1.37	-2.27	-2.44
Likelihood Chi-square	537.4	482.0	413.4
Degrees of Freedom	550	543	534
Percentage of Baseline Chi-square Explained by Model	14.9	23.6	34.5

(1) For dichotomous variables, the coefficient is for the category shown in parentheses. The coefficient for the other category is the negative of this coefficient. For variables with more than two categories, coefficients are shown for all categories. When there are two or more categories, the significance test is done for all categories together.

*Significant at .1, **significant at .05, ***significant at .01

significant effect on migration once the other variables are taken into account. All of the effects of these variables should be indirect through their relationship with the social and economic bonds and the intervening variables. Unfortunately, this prediction is not supported by the data. While there is some reduction in the effects of age and urban/rural origin when social and economic bonds and the intervening variables are entered, the effects of sex and education increase, as judged by the size of the coefficients.[4]

Sex has little effect on migration when only the background variables are included in the model, but the effect becomes significant when the social and economic bands are added. This suggests that some of these bonds act differently for men and women. We shall examine this question later by separating the analysis by sex.

Age has its normal relationship to migration. At ages under 30, the coefficients are positive indicating that the ratio of movers to stayers is above average for these groups while at ages 30 and over, the coefficients are negative.

Education retains its strong positive effect on migration when the other variables are controlled. This means that persons with higher levels of education are much more likely to move than those with lower levels. Apparently only a small part of this relationship is due to the larger awareness space, greater job information and higher income expectations of those who are highly educated. One motive for the migration of highly educated persons is to obtain additional education, which is available only in cities. This is particularly true for young persons who are senior graduates who wish to attend a college or university. Approximately ten percent of the migrants said they moved to attend school. A large proportion of all colleges and universities in Taiwan are located in Taipei.

The positive coefficient for urban/rural origin indicates that migration from urban areas to Taipei is somewhat greater than from rural areas. However, the effect is relatively weak when the intervening variables are added to the model. This suggests that some of the higher migration from urban areas is due to the greater amount of job information and greater awareness of opportunities elsewhere among urban residents compared to rural residents.

Among the six measures of social and economic bonds tested, five had significant correlations with migration. These were marital status, whether or not the respondent decided household finances, whether or not the family

----

[4] The effects of a particular variable on migration can be estimated by taking the exponential of the coefficient and interpreting the result as a multiple of the ratio of movers to stayers. For example, in model C the coefficient for sex is .332. The exponential of this number is 1.39. This means that being male increases the ratio of movers to stayers by 1.39. Females have a multiplier which is the reciprocal of this, or .72.

owned their housing unit, whether the family owned land and whether or not the person was self-employed. Those who had financial control and those who owned a house or more than 2 hectares of land, or had their own business were less likely to move than those who did not. Being married had an unexpected positive effect on migration in the multivariate analysis. This result was unexpected because it had been predicted that being married would reduce one's chances of moving. We shall see later that this relationship can be attributed to the females and that marriage appears to make little difference for males.

The last potential bond, having a job at the place of origin, has only a small effect on migration. While those who were previously employed are less likely to move, the effect is not statistically significant.

Turning to the intervening variables we see that only three of these have significant independent effects on migration. These are the two cost-benefits variables: the expected increase in income and the expected cost of moving, along with the possession of job information. Overall, the expected cost of moving has the largest coefficient of any of the intervening variables in the regression, indicating that differences in costs arising from differences in household composition and possessions to be transported along with transaction costs associated with moving may partly answer the question of why one person choses to move in response to an opportunity elsewhere while another does not move. It is also possible, since the question refers to perceived cost, that migrants have more accurate perceptions of costs than nonmigrants.

In the preceding analysis we have dealt with a rather heterogeneous group which contains men and women and persons who are in and out of the labor force. The perceived costs and benefits of moving and many of the other determinants of migration may differ widely between these groups. A better fit of the model to the data is likely to be achieved by studying various sub-groups separately. In Table 6.5, logistic regressions for three main sub-groups are shown: males in the labor force, females in the labor force, and females not in the labor force. Labor force status was determined at the time of interview. Therefore, it must be assumed that those in the labor force then were also in the labor force at the time of moving or that they were interested in entering the labor force soon after they moved and that the desire to do so played an important role in their decision to move. We have not included men who were not in the labor force because there were only 18, too small a group for separate analysis.

None of the variables has consistent relationships which are statistically significant for all three sub-groups. However, whether the respondent is the household decision-maker has a consistent positive relationship which is significant for two of the groups and cost of moving and homeownership have consistent negative relationships in all three groups, but the coefficients are not significant for females in the labor force. Marital status has a significant positive effect for women, indicating that female migration is enhanced through marriage. Employment at the place of origin and self employment deter mobility for those who are in the labor

TABLE 6.5

LOGISTIC REGRESSION COEFFICIENTS FOR DETERMINANTS OF MIGRATION BY SEX AND LABOR FORCE STATUS

	Males in Labor Force	Females in Labor Force	Females not in Labor Force
A. Background Characteristics			
Age at Move: 15-24	.937**	-2.480*	.586
25-29	.907	.585	.155
30-34	-1.136	.874	.023
35-39	-.708	1.021	-.764
Education: Less than 5 years	-.117	.766	-1.154**
Primary Grad	-.459	-.140	-.628
Junior High	-.272	.383	.043
Senior High	.848	-1.009	1.739
Urban/Rural Origin (Urban)	.096	-.269	.550**
B. Social and Economic Bonds			
Marital Status at Move (Married)	.514	.976*	1.241***
Decision Maker (R or Spouse)	-.355	-2.145***	-.786***
Homeowner (Yes)	-.586**	-.522	-.849***
Land Owned (Any)	-.132	-1.146**	.043
Job at Origin (Yes)	-1.284***	-1.747**	-
Self-employed (Yes)	-.852***	-1.916*	-
C. Intervening Variables			
Residential Satisfaction (Not Satisfied)	-.474**	1.314***	.395*
Job Satisfaction: Satisfied	.489	-.063	-
Indifferent	-.015	.517	-
Dissatisfied	-.357	-.454	-
Awareness Space (3+ cities)	-.186	.934**	.191
Job Information (Yes)	.370	1.785***	-
Living Conditions Information (Yes)	.132	-.775*	.289
Expected Increase in Income (Yes)	.799***	1.296***	-
Expected Moving Costs (Yes)	-1.014***	-.118	-.627***
Constant	-.692	-.670	-2.020
Likelihood Chi-square	132.5	44.8	141.9
Degrees of Freedom	219	85	177
Percentage of Baseline Chi-square Explained by Model	49.5	66.2	29.5

See notes to Table 6.4.

force, implying that migration rates are higher for those who are unemployed or entering the labor force and for those who are employed by others.

Among the intervening variables, residential satisfaction, which had no effect for the overall group, has significant but opposite effects in the different sub-groups. It has the predicted effect (higher mobility with greater dissatisfaction) for females but the reverse effect for males. The expected increase in income is important for males and females in the labor force. Job information and awareness space are important for females in the labor force, but not for males. Information on living conditions has an unexpected negative effect for women in the labor force.

When these other variables are controlled, the background variables have relatively little effect on migration within the three groups. Age has the expected negative effect for males in the labor force and for females not in the labor force, although the relationship is not significant in the latter group. For females in the labor force, migration increases with age. Education and urban/rural origin have significant effects on migration only for females in the labor force.

Overall, the results presented in Table 6.5 make it clear that different models should be used for males and females and for persons in and out of the labor force. The level of explanation for the two groups that are in the labor force is considerably higher than that for the total population (as indicated by the percentage of baseline chi-square explained by the model). The fact that the level of explanation is lower for those who are not in the labor force suggests that the model and particular set of variables chosen for this study apply best for labor migration and that further work is needed to find a better model for the migration of persons who are not in the labor force.

CONCLUSION

The main purpose of this chapter has been to develop and test a theory of individual and household migration. The human capital approach with its emphasis on the costs and benefits of migration was found to be incomplete for several reasons. One of the main objections to the human capital approach, which was raised in the first section of the chapter, was its reliance on the assumption that everyone was constantly reassessing the optimality of their present location and considering whether or not they would be better off if they moved to another location. This assumption is clearly contradicted by observations of actual behavior. For example, in the 1973 survey in Taiwan, only nine percent of the residents outside Taipei reported having considered a move to Taipei at any time during the previous five years and only an additional eight percent had considered any move.

It was also argued that the human capital approach did not take account of the major role of friends and relatives in determining who decided to move and where they decided to move. In addition, the role of information and the awareness of opportunities elsewhere was neglected in most previous applications of the human capital approach. In an attempt to correct these weaknesses of the human capital approach, a more general model was developed in which costs and benefits were only one of four sets of factors involved in determining migration. The other three factors were: economic amd social bonds, satisfaction with the present location, and awareness of opportunities elsewhere. These additional factors helped to explain why most nonmigrants had not considered moving. Non-migrants were hypothesized to have more social and economic bonds to their present place of residence, greater satisfaction with that place, and less awareness of opportunities elsewhere than migrants.

The extent to which the four factors in the model influence migration was investigated for a sample of migrants to Taipei City and a comparison sample of residents in urban and rural areas outside Taipei. Both samples were restricted to persons aged 20 to 39 at the time of interview. Most of the comparisons were between the migrants' situation before the move (as reported in the interview after the move) and the residents' situation at the time of interview. There was a possibility of bias due to changes in overall conditions between the period prior to the move and the time of the interview and also due to recall lapse; however, these biases were minimized by restricting the migrant group to "recent migrants" who were defined as having moved within the past five years so that the average time lapse between the time of the move and the interview was around two and one-half years.

Several social and economic bonds were examined. The proportion of migrants who were living with their parents prior to moving did not differ significantly from the proportion of residents living with parents when age was controlled. However, more migrants than residents were living in households where they had little say in the expenditure of household income, so that freedom to spend one's own income may have been a factor influencing migration. In this sense, the effect of the bond is opposite to that of other bonds. Clearly the nature of the bond is important and bonds which place people in subordinate positions may be negatively evaluated and thus provide a reason for moving.

Economic bonds also varied in their relationship to mobility. Migrants were less likely to have been employed or to have had a farm job, if employed, before moving than stayers. Migrants were also less likely to have been homeowners than stayers. However, there was little difference in the ownership of farmland among the two groups except that migrants were less likely than residents to own large farms of two or more hectares. Apparently the smaller farms were being farmed adequately by other family members.

Our measure of subjective residential satisfaction, which consisted of housing and neighborhood items plus distance to work and the quality of local schools, failed to distinguish migrants from residents. Similarly, job satisfaction did not differ significantly between migrants and residents.

The failure of these satisfaction measures may have been due in part to distortions in the recall of these items for migrants. These subjective measures are probably more likely to have been biased by retrospective measurement than the more objective items such as home ownership and previous living arrangements. Responses to questions on satisfaction may also have been biased by norms of politeness toward strangers (interviewers) which are particularly strong in the Chinese culture.

Awareness space proved to be a marginally useful concept in this study. Nearly one-half of the residents had a general familiarity with Taipei and one-quarter viewed it as a favorable place to live. Yet none of them moved. The average number of urban places in the awareness spaces of migrants and residents was between three and four and there were no consistent differences between the two groups. However, migrants were more likely than residents to have received specific job information from friends and relatives. Thus, specific knowledge of opportunities rather than general awareness appears to be important in the second stage of the decision-making process.

An analysis of expected costs and benefits of moving indicates that many more migrants than residents had expected the move to result in an increase in income. Migrants also experienced significantly lower moving costs on the average than residents expected to pay if they moved. Not all migrants moved for economic reasons and other measures are needed to determine what benefits these migrants expected to receive from moving.

Finally, the multivariate analysis showed that the total set of bonds and intervening variables increased the explained variance in migration substantially from that which was explained on the basis of background characteristics alone. This provides support for the main elements of the model.

7

The Economic Success of Migrants in Taipei

City-ward migrants, especially those from rural places and small towns, are usually distinct from lifetime city residents in many aspects. At the place of destination, the newcomers need to adjust to the environmental changes and to realize their expectations of the movement as much as possible. Previous studies have examined the individual outcomes of migration to big cities along several different dimensions (for a review, see Tirasawat, 1977: 31-45). Focusing on the case of Taipei, this research limits its scope to an intensive study of the economic perspective of migration consequences. There are two underlying grounds for this limitation. First, as in many other cases, the prevailing reported reasons for moving to Taipei were job-related. Second, Taiwan is a small island with a well-developed transportation network and the majority of migrants to Taipei are members of the dominant ethnic group in the city. There appears to be little social and cultural gap or language problems between the place of destination and the areas of origin. Hence, other dimensions of migration consequences are relatively much less important.

The migration literature has provided strong empirical support for the importance of economic incentives in the decision to migrate (Shaw, 1975; Somers, 1976; Todaro, 1976a; U.S. Bureau of the Census, 1977; Yap, 1975). Studies on the determinants of migration mostly indicate the tendency for people to move to places with higher expected opportunities. Some investigations further reveal that migrants are actually better off after moving to big cities (for a review, see Tsay, 1979: 19-27). In other cases, however, city-ward migration is thought to add substantially to the low-wage, low-productivity work force in the urban traditional sector. Subsequently, reduction and redirection of migration flows are among the policy options suggested for alleviating urban poverty and unemployment (Frank, 1968; Peek and Antolinez, 1977). This suggestion might have overlooked the important relationship between migration and the urban labor market. Migrants' experience in changes of their economic characteristics due to migration should be taken into consideration.

This chapter attempts to evaluate the individual outcomes of migration to Taipei by analyzing net differences in economic characteristics (with concentration on employment and earnings) between migrants and otherwise similar nonmigrants in places of origin and in the receiving area. The major purposes are (1) to ascertain the effects of migration on migrants, (2) to explore the sources of migrant gains, and (3) to assess whether migrants are well assimilated into the urban labor market of Taipei. Hopefully, the analyses will provide some insight for policies intended to modify migration patterns and city population configurations.

RESEARCH DESIGN

This research aims to answer three questions which are most relevant to the study of individual outcomes of city-ward migration. (1) Are migrants better off after moving to the city in terms of earnings and employment characteristics than they would have been had they stayed in the sending areas? (2) If migrants are better off, what are the sources of their improvement? Have migrants benefitted simply because they have moved to a city which has experienced more favorable changes in labor market conditions than their place of origin? Or, is the improvement partly due to changes in the migrants themselves after movement? (3) If part of the improvement is the result of changes in migrants, what is the process and nature of these changes? The process of such changes in migrants is the economic dimension of migrant adjustment (Boulier, 1977). Ideally, to answer the questions raised above would require a longitudinal study design in which migrants are followed from the time of their decision to move through settlement in the city. Unfortunately, only cross-sectional survey data collected at a single point in time are available. Therefore, in accordance with the study objectives, this research has three phases involving comparisons of: (1) migrants versus otherwise similar stayers at places of origin, (2) migrants versus otherwise similar city natives, and (3) migrants at arrival in the city versus migrants at the time of interview.

Assuming that migration is undertaken primarily for economic reasons as most migrants report when asked for their reason for moving, it is possible to ascertain the effects of migration, at least in part, by measuring income differentials of migrants before and after moving. However, this measure has some basic problems. First of all, the increase in income needs to be adjusted for the effect of overall economic growth. Such an adjustment requires information about income changes of nonmigrants (or of the whole country) over the same period of time. Because this period of time varies from migrant to migrant, it is empirically difficult (if not impossible) to define the time period of reference for nonmigrants and to measure their income changes. Secondly, the income difference needs to be deflated by the difference in price level between area of origin before migration and area of destination after migration. This would be a laborious task because migrants have moved from various places of origin at different points in time. Finally, except for longitudinal studies, the characteristics of migrants prior to moving are obtained after migration has occurred. The retrospective information may be distorted by recall lapse, especially in the cases of long-term migrants. In addition, migrants may attempt to make their economic positions before migration appear more (or less) in line with their expectations and outcomes of the movement than they actually were.

An alternative approach, which is the first one followed in this chapter, is to examine the effect of migration by comparing the current incomes (at the time of interview) of those who have moved with the current incomes of those otherwise similar individuals who have not.[1] Such a comparison is made under the assumption that the incomes of stayers in the sending area are accurate estimates of the incomes which migrants would have received if they had remained at the origin. To the extent that this assumption holds, this method solves two problems. First, because comparison is made at the same point in time, the effect of overall economic growth is a constant and income needs only to be adjusted for the current differential in cost-of-living between area of origin and area of destination. Second, current data are much more reliable than retrospective information. However, it is difficult to control all the background characteristics which affect income levels so that migrants differ from stayers only to the extent that the former have moved but the latter have not. Therefore, the income difference as measured by this method would contain some residual effect of migrant selectivity, in addition to the effect of changes in migrants themselves and that of favorable changes in the urban market over the places of origin.

In addition, there are some other restrictions to this method. (1) This research focuses only on private costs and returns; social costs and returns will not be considered. Moreover, our main interest is the economic benefits of migration rather than the costs of movement which are essentially noneconomic (Somers, 1976). (2) Returns to the investment in migration are probably confounded with returns to other human capital investments on the part of migrants. Whereas the effect of education is relatively easy to control, the gains resulting from investments in training, labor market information, and job search time and expenditures are difficult to identify and are likely to increase the economic benefits associated with migration. (3) This research does not take into account psychic returns to migration which are not covered by the available data. (4) The effect of migration would be somewhat inflated (deflated) if return and onward migrants are negatively (positively) selected with respect to economic success in the city.

----

[1] This idea was suggested by Somers (1967) and is similar to the cost-benefit model (Sjaastad, 1962; Speare, 1971a) of migration decisions, except that the cost factor is ignored and the outcomes of migration, instead of the expectations from migration, are considered.

The results of the first phase of this research may show that migrants are significantly better off after moving. The migrant gains might result from two major factors. One is the effect of economic adjustment made by the movers. The other pertains to the more favorable market conditions in Taipei than in places of origin.[2] To distinguish the effect of adjustment from that of other general changes, the second phase will be to compare the characteristics of migrants at the time of interview with those of otherwise similar city-born residents. As we know, the urban labor market conditions affect both migrants and urbanites. If there is no significant difference in economic placement between the two groups, the results favoring migrants over stayers who remained in places of origin must be mainly due to labor market changes which favor the receiving area over the sending area. However, if migrants are economically inferior to otherwise similar lifetime city residents, this difference will reveal the disadvantages of migrants.

It is hypothesized that migrants from urban areas[3] should be as well off as Taipei natives because urban-to-Taipei migrants have relatively few disadvantages associated with their place of origin. On the contrary, migrants from rural areas are not expected to be as economically successful as their city-born counterparts. Again, the duration of residence of migrants in the city is considered crucial to their economic success.

To test the hypotheses, a composite earnings function will be estimated for all individuals in the urban labor force, including both Taipei natives and migrants. The earnings determinants will include human capital variables such as education and experience as well as a set of binary variables representing migration status. There are five categories of migration status as follows: (1) Taipei natives (including those who moved to Taipei before 15 years old), (2) long-term urban-to-Taipei migrants, (3) recent urban-to-Taipei migrants, (4) long-term rural-to-Taipei migrants, and (5) recent rural-to-Taipei migrants. Long-term migrants refer to those who moved to Taipei more than five years ago. Recent migrants are those who migrated to Taipei in the past five years. Using Taipei natives as the reference group, the estimated coefficient of a dummy variable representing migration status shows the percentage deviation in earnings of

---

2 Migrant gains will surely be exaggerated to the extent that the negative selectivity (in terms of economic success) of return and onward migrants and the positive selectivity of migrants (as compared to stayers) are not able to be quantified and controlled (Tsay, 1979: 39-44). A comparison of in and out-migrants made by Speare (1974a) suggests that, if anything, out migrants from major cities are positively selected. Thus, they may not be a problem.

3 All cities and townships with 50,000 population and over are classified here as urban areas. All other townships are rural areas (Tsay, 1979:33-34).

the individuals characterized by that variable from that of Taipei natives. If long-term migrants score higher on the measures of success than recent migrants when background variables are controlled, it will be inferred that adjustment has taken place. In addition, the hypotheses will be tested in terms of employment by means of an analysis of the labor mobility of migrants as compared to that of city natives.

Finally, the third phase of this research will involve further investigation of the changes in migrants' economic status and behavior over time after their arrival in the city. Ideally, the study of the adjustment process requires longitudinal observations on the changes in characteristics of migrants and their behavior (Boulier, 1978; McCutcheon, 1978; Speare and Goldstein, 1978). Since such ideal data are not available, this study examines the retrospective histories of migrants in terms of job search and occupational mobility. It is understood that there are several problems in such an examination since (1) those migrants who remained in the city may have been more or less successful than those who left through return and onward migration, and (2) changes in characteristics of migrants may be associated with age and/or general changes in the city (Boulier, 1977; Zimmer, 1973). However, the effects of these two processes might be controlled in part by holding constant those characteristics most related to the propensity to move.

According to the two-sector analysis of labor migration and urban unemployment (Harris and Todaro, 1970; Todaro, 1969), a migrant moves to a city to seek a modern job. Since migration decisions are based on expected rather than actual gains from migration, accelerated rural-urban migration may coexist with rising urban unemployment in the cities of less developed countries. In the two-sector model, traditional sector employment in the cities is inappropriately treated as being equivalent to unemployment and the number of unemployed persons is obviously over-expected (Fields, 1975; Speare, 1977). In reality, a migrant may pursue one of the following two strategies after arrival in a city: (1) he may remain unemployed until he either obtains a modern job or loses hope and returns to his place of origin, or (2) he may take a temporary job in the urban traditional sector while searching for a modern sector job (Anderson, 1977; Fields, 1975).

This theory applies primarily to male migration because most males, and especially those in the age group of 20-39 considered here, move for job-related reasons. While some females move for job-related reasons, a large proportion give family and other reasons. This study will investigate whether migrants to Taipei acted in the rational manner assumed by these theories.

THE COMPARISON BETWEEN MIGRANTS AND STAYERS

The present section is an empirical evaluation of the effect of migration on individual migrants. At the micro-level, this research will assess the net differences in economic characteristics between migrants in Taipei and stayers in the sending areas. The main focus of the investigation will be placed upon migrant gain in terms of actual increase in earnings which have been adjusted for the cost-of-living differences between the receiving area and the places of origin (for details see Tsay, 1979: 51-60). After examining the income benefits, attempts will also be made to ascertain the migration effects by comparing the employment characteristics of urban-to-Taipei migrants with those of the residents in other cities and big urban towns.

Earnings of Migrants versus Stayers

There are both theoretical grounds and empirical evidence for expecting that migrants will experience gains in income. However, the actual outcome of a move may not necessarily follow what a migrant anticipated because migration involves some element of risk. This is very much the case in Taiwan where the majority of movers do not have a prearranged job and where migrants receive information about job opportunities in the city mainly from informal channels, such as friends and relatives (Speare, 1973). Moreover, market opportunities are usually not stable over time, especially with rapid increases in the urban labor force.

In the initial period of stay in the city, migrants are likely to experience little income benefits from moving if they must accept casual, intermittent employment before finding a better job. With increasing length of residence, some migrants may continue to experience only small income gains or even losses in income if they are unable to compete with the city-born. In the urban labor market, the newcomers may be handicapped by such factors as a lower quality education, less accumulated urban job experience, and job discrimination from employers. Nevertheless, long-run gains are likely to be observed among migrants, due to their assimilation in the urban labor market and to the tendency for disappointed movers to return home or move on.

The sample used for the comparisons of earnings is limited to males who worked in nonfarm employment or who worked on farms for wages. Farmers working on their own family farm and unpaid nonfarm family workers[4] were not asked about their earnings. Because the proportion employed in agriculture in Taipei is trivial, it is believed that the comparisons based on nonagricultural earnings alone can serve the purpose of evaluating the effect of the city-ward migration on migrants. In fact, the discrepancies revealed by such comparisons are conservative estimates of the actual differences in over-all earnings between Taipei and the sending areas (especially the rural part) where the share of males employed in

----

[4] Less than 1 percent of the males in the sample were unpaid family workers in the nonagricultural sector.

agricultural is relatively much more substantial.[5]

Initially, this study will assess the effect of migration on migrants by comparing median earnings between movers and stayers. Afterwards, the assessment will be carried out by using dummy variable regressions to estimate and contrast earnings functions of migrants and nonmigrants. Since duration of residence in the city has been found to have positive effects on migrants' earnings (e.g., Oberai, 1977; Peek and Antolinez, 1977; Yap, 1976), migrants are further dichotomized into recent and long-term categories. In computing all the statistics, a weighting procedure is used wherever appropriate because of the heavier sampling of recent migrants and married persons[6]. In performing tests of statistical significance, however, the total weighted sample size is adjusted downward to approximate the actual sample size, so that the weighting will not greatly affect the significance tests.

The earnings of migrants and nonmigrants at the place of origin are shown in Table 7.1. Medians rather than means are used because the distributions are skewed to the right and mean values are heavily influenced by a few cases with extremely high earnings. This influence can be particularly strong when the sample size is small. Since the median is free from the heavy influence of extreme values, it is superior to the mean as a measure of central tendency of the earnings distributions.

Comparisons of median earnings support the hypothesis and reinforce the earlier conclusion that migrants have greater earnings than stayers. This relationship persists even when controlling for place of origin. The median income of migrants from both urban places and rural areas exceeds that of stayers in the corresponding category. Furthermore, the relationship holds unchanged for migrants of any length of residence in Taipei. Consistent with the research hypothesis, the income difference is

----

[5] For instance, according to Liu (1974: 169-179), the 1969-1972 average rate of return to labor per employed person in the agricultural sector is only one-third of that in the nonagricultural sector. According to the survey data used in this study, the proportion of male residents aged 20-39 employed in the agricultural sector is 1.4 percent in Taipei City, 21 percent in the urban areas outside Taipei, and 54 percent in rural Taiwan. No migrant reported working on farms.

[6] In the 1973 Taiwan Migration Survey, married males had twice the probability of being included as single males due to the possible selection of wives as respondents. In Taipei City, recent migrants were sampled at two times the rate of the other respondents. The weights for various groups of migration-marital status are as follows: (i) 1 for married recent migrants and married stayers, (ii) 2 for single migrants, single stayers, and married residents in Taipei other than recent migrants, and (iii) 4 for single residents in Taipei other than recent migrants.

TABLE 7.1

MEDIAN MONTHLY EARNINGS OF MIGRANTS AND STAYERS BY EDUCATION AND PLACE OF ORIGIN

Place of Origin and Education	Migrants		Non-migrant at Origin
	Recent	Longterm	
All Educational Levels			
All Origins	3860	4255	3208
Urban Origins	3878	4266	3704
Rural Origins	3874	4134	3205
More Educated - Junior High Graduates and Above			
All Origins	4238	4743	4052
Urban Origins	4303	4761	4052
Rural Origins	3958	4240	3840
Less Educated - Less than Junior High Graduates			
All Origins	3362	3736	3079
Urban Origins	3377	3710	2916
Rural Origins	3070	3909	3164

Source: 1973 Taiwan Migration Survey.

Notes: 1. All earnings data have been adjusted for cost-of-living differentials and are expressed at the cost-of-living level in the destination of Taipei City (Tsay, 1979:51-60).

2. Income measured in Taiwan dollars (1 U.S. $ = NT S38 in 1973).

especially sharp for long-term migrants in comparison with recent migrants (NT$ 1,047 vs 652). This positive effect of duration of stay in the city upon migrant gains is evident for migrants of both urban and rural origins. For those of urban origin, the median earned income of long-term migrants is NT$ 562 higher than that of stayers, while for those of rural origin, the long-term migrants earned NT$ 929 more. There is also a difference in earned income between those who recently moved and nonmigrants in both urban and rural sending areas.

The fact that migrants, regardless of duration of stay in the city, have higher earnings than nonmigrants in places of origin may be due to two major reasons. First, the labor market conditions are better in Taipei than in the sending areas. Consequently, the probability of obtaining high-wage jobs is greater for migrants than for those who remain behind. Secondly, as has been found in many studies (e.g., Barnum and Sabot, 1976; Caldwell, 1968, 1969; Herrick, 1965; Oberai, 1977; Speare, 1973, 1974a), migrants to Taipei are positively selective of more educated persons. The proportion of migrants in the category of junior high school graduates and over is 59 percent while the proportion for stayers is only 35 percent. Since education is an important factor in determining earning capacity (Merrick, 1976; Mincer, 1974; Oberai, 1977; Tandon, 1978), this variable should be held constant in order to assure that the relationship between geographic mobility and labor income is not due simply to differences in education.

Table 7.1 also presents separate comparisons of monthly earnings between migrants and stayers for the more educated (junior high school graduates and over) and the less educated (less than junior high school graduates). As hypothesized, after controlling for education, the average earned income of migrants remains higher than that of stayers. For the more educated category, long-term migrants have a median income of NT$ 4,743 as compared to NT$ 4,053 for stayers. A similar pattern is observed among those with less education (3,736 versus 3,079). Recent migrants also have higher earnings with the exception of less educated migrants from rural origins.

The figures in Table 7.1 further suggest that duration of residence in Taipei clearly exhibits a positive influence upon migrants' median earnings even when education is controlled. For both the more educated and the less educated groups, the income gain made by long-term migrants consistently exceeds that made by recent migrants in all rural-urban classifications. Although less educated migrants from rural origins did not do quite as well in the short run as stayers they did significantly better in the long run.

In short, the direct comparisons reveal that migrants in most cases have achieved significant gains in earnings as compared to stayers. This relationship persists even after education, place of origin, and/or duration of residence in the city are controlled. The findings support the research hypothesis that the movers are better off due to migration. However, it is worthwhile to note that the direct comparisons are deficient due to the lack of detailed and simultaneous controls for earnings determinants. For example, the division of educational levels into the more educated and the less educated at the line of junior high school graduation is crude and may have done only a partial job of controlling for education. The observed discrepancies in earnings between migrants and stayers may be attributable to the uncontrolled portion of the educational differentials between the two groups. For example, among the more educated from urban origins, recent migrants had 13.6 years of education, long-term migrants had 12.9 years and stayers had 11.9 years. A significant difference also exists between the less educated rural stayers and long-term migrants. Therefore, it is advisable to employ a more elaborate technique such as regression analysis to perform more detailed and simultaneous controls for earnings determinants such as education.

Regression Analyses

In the framework of regression analyses, current earnings of migrants will be compared with their foregone earnings, which will be estimated as the current average earnings of otherwise similar nonmigrants who remained behind. Since simultaneous controls for earnings determinants are possible in regression analyses, differences in earnings attributable to migration status can be distinguished from differences due to the unequal amount of human capital. This study anticipates that the regression coefficient of each dummy variable representing migration status (with the stayers as the omitted reference category) is significantly greater than zero.

The specification of the earnings functions in this study is guided by the human capital theory (Becker, 1964) which provides a useful basis for empirical studies in identifying major determinants of individual earnings. The human capital model was developed and applied to schooling by Becker and Chiswick (1966). Mincer (1974) extended it to include post-schooling training. The human capital models take the length of schooling, level of training, and experience as the basic cause of the inter-personal variation in labor income. In this research, the human capital formulation of earnings functions is augmented with a set of categorical dummy variables representing migration status. The earnings functions are of the following form:

$$LnY = a + bS(1) + cS(2) + d(ED) + e(EP) + fSQR(EP) + D.$$

where:

LnY = natural log of monthly earnings of an individual,[7]

----

[7] The earnings distributions examined in this research are quite skewed to the right. In order to reduce the degree of dispersion, monthly earnings

S(i) = migration status i of the individual (stayer (omitted), recent migrant (i=1), long-term migrant (i=2)),
ED = years of schooling of the individual,
EP = potential years of work experience of the individual,
SQR(EP) = the potential years of work experience squared,
D = the disturbance term which includes the effect of natural ability and luck.

The coefficients of these dummy variables measure the partial effects (on the dependent variable) of being in one category relative to the omitted category of stayers. For continuous explanatory variables in the semi-log regressions, the regression coefficients show the relationships in relative change between the variables and the monthly earnings. For example, one year of change in schooling would cause changes in earnings of about 100*d percent.[8] That is, the average rate of return to schooling is 100*d percent.

With regard to potential years of work experience, this research adopts the estimator used in Tandon's (1978) study on earnings differentials. In accordance with the post-schooling investment models (Mincer, 1974), potential work experience is set equal to age, minus years of education, minus six (the number of years before attending school). For people having the same number of years of schooling, the older they are, the more experience they have. However, for those with the same age, the ones with more education have less experience than those with less education. This investigation hypothesizes that the regression coefficient of years of experience squared is negative. In other words, earnings increase with experience at a more rapid rate in early years than later years.

Separate regressions are run for rural and urban areas of origin, so that the current labor income of migrants can be compared with that of nonmigrants in the equally urbanized places of origin. Each regression includes both those who still remained in the area (i.e., stayers) and those who moved from the area to Taipei (i.e., migrants). Each observation is weighted to compensate for the fact that married males had twice the chance of being included as single males.

The regression estimates reported in Table 7.2 describe a familiar earnings pattern. The returns to education are strongly positive. To both the urban group and the rural group, the coefficient for schooling is about .051. That is, on average, an increase/decrease in schooling by one year would cause an increase/decrease in monthly earnings by 5.1 percent. The

---

are transformed to natural logarithm values before entering them into the regressions. This transformation is widely recommended and commonly practiced among income-related studies (Merrick, 1976; Mincer, 1974; Tandon, 1978; Yap, 1976).

[8] This is true when d is very small. For larger values of d, the percentage increase is 100[exp(d)-1].

TABLE 7.2

EARNINGS AS A FUNCTION OF MIGRATION TO TAIPEI, 1973

Independent Variables	Regression Coefficients Total (1)	Rural (2)	Urban (3)	Total (4)
Dependent Variable: Ln (monthly earnings in N.T.$)				
Migration Status				
Recent migrant	0.1594** (2.358)	0.2754** (2.142)	0.0824 (1.045)	
Longterm migrant	0.1810*** (2.817)	0.2365** (2.246)	0.1395* (1.701)	
Rural migrant				0.1986*** (2.949)
Urban migrant				0.1480** (2.346)
Years of Schooling	0.0511*** (5.785)	0.0505*** (2.938)	0.0514*** (4.988)	0.0522*** (6.000)
Potential Years of Work Experience	0.0342** (2.208)	0.0033 (0.100)	0.0505*** (2.923)	0.0349** (2.320)
Potential Years of Work Experience Squared	-0.0008* (1.703)	0.0002 (0.253)	-0.0013** (2.516)	-0.0008* (1.744)
Constant	7.3941	7.5419	7.3370	7.3768
R Square	0.135	0.131	0.156	0.136
Sample Size	386	162	224	386

Notes: The t-ratio is in parentheses. ***Statistically significant at 1 percent level for two-tailed t-test. **Statistically significant at 5 percent level for two-tailed t-test. *Statistically significant at 10 percent level for two-tailed t-test.

coefficients of the experience variables are also positive, but much stronger for the urban group than for the rural group. The coefficient for the square of years of work experience is significant with the expected negative sign for the urban group, but not significant for the rural group. This means that in urban areas earnings increase with years of experience in a curvilinear fashion, while in rural areas there is little increase in earnings with years of experience.

Despite the rapid population growth in Taipei in recent years, Table 7.2 shows significant income gains associated with migration to the city from both urban and rural places. In the rural regression, the gains are visible shortly after moving as indicated by the significant coefficient for recent migrants. In the urban comparison, the initial increase in earnings is small and not statistically significant. After five years or more in the city, however, migrants of either rural or urban origin earn significantly more than the corresponding stayers controlling for the amount of human capital.

The magnitudes of the migrant gains are substantial. Regression (3) in Table 7.2 shows that recent urban-to-Taipei migrants have earnings about 9 percent (i.e., exp(0.0824)-1) more than urban stayers, holding length of schooling and experience constant. With increasing duration of residence in Taipei, the gain amounts to 15 percent (i.e., exp(0.1395)-1) for long-term migrants of urban origin. Greater income gains are discovered in the rural regression. For recent movers from rural places, the average level of earnings is 32 percent (exp(0.2754)-1) higher than that of nonmigrants in the countryside. Long-term rural migrants enjoy earnings 1.27 (exp(0.2365)) times that of rural stayers. The observed pattern that recent rural migrants have higher earnings than their long-term counterparts is probably due to the few extreme cases in the small sample of recent migrants. The average earnings gains accruing to all migrants in Taipei are estimated at about 20 percent[9] of the average labor income they would have received had they remained in the sending areas.[10]

----

9 Derived from taking the average of the magnitudes of migrant gains for all groups of migrants.

10 As expected, the amount of migrant gains assessed by this research is not as large as the 50 percent figure reported by Yap (1972, 1976) for Brazil. Such a disparity perhaps is due to two major factors. First, cost-of-living adjustments were not made in Yap's investigation. Second, in assessing migrant gains, the Brazilian study was limited to the comparison between rural-to-urban migrants and rural stayers. In the Taipei study, the gain is 22 percent when only rural-to-Taipei migrants are compared with rural stayers (see Regression (4) in Table 7.2). Since the cost-of-living in Taipei is estimated at 28 percent higher than that in rural Taiwan (Tsay, 1979: 58-60), the migrant

It should be acknowledged, however, that the estimates of net gains to migrants were likely affected by the adjustment made for cost-of-living differentials between migrants and stayers. The earnings of stayers in rural areas, small urban places, and provincial cities were inflated by 28 percent, 16 percent, and 6 percent, respectively, before they were compared with the earnings of migrants in Taipei (Tsay, 1979: 58-60). It is clear that, to the extent the cost-of-living differentials are overestimated (underestimated), the net gains to migrants are underestimated (overestimated). A more complete set of data for estimating the discrepancies in cost-of-living between the sending and receiving areas would undoubtedly result in a more precise estimation of the magnitudes of migrant gains.

It should also be repeated here that the migrant gains assessed in this research are based on comparisons of nonagricultural earnings between migrants and stayers. Because earned income is significantly lower in the agricultural sector than in the nonagricultural sector (Liu, 1974: 169-179), the estimated income gains accruing to migrants would have been even larger if the earnings of those stayers employed in the agricultural sector had been taken into account.

There are several possible reasons for the higher earnings of migrants. Earnings in Taipei may be higher than those in the rest of Taiwan. Higher migrant earnings may also be due to a higher return to education and experience in Taipei than in places of origin. However, when interaction terms designed to pick up these differences were added to the regression, they had insignificant effects. Thus, they are not included in the final regressions reported in Table 7.2. The lack of significance may be due to multicollinearity and small cell sizes.

Finally, the average earnings of stayers may underestimate a migrant's foregone earnings if migration is positively selective in terms of characteristics other than years of schooling and potential work experience (Heckman, 1979; Nakosteen and Zimmer, 1980). Actual migrant gains will also be exaggerated if return and onward migration are selective of the economic failures. While little is known about the extent of these possible upward biases, there are downward biases which may offset them. For example, migrants' foregone earnings will be overestimated to the extent that migrants had higher unemployment rates prior to moving. Moreover, for migrants who increased their skill level by obtaining more formal education or on-the-job training in Taipei, the estimated earnings gains are conservative. Therefore, correcting them for all these biases, if it were feasible to do so, would not be likely to erase the gains associated with the city-ward migration.

----

gains would be about 50 percent had the cost-of-living differentials not been adjusted.

TABLE 7.3

LABOR FORCE STATUS OF MALE MIGRANTS AND NON-MIGRANTS AT PLACE OF ORIGIN

	Migrants		Non-migrants	
	Recent	Long-term	Urban Areas	Rural Areas
Working	76.9	92.1	93.7	96.9
Unemployed	9.7	1.6	2.5	0.8
Student	11.2	6.3	3.0	0.0
Other	2.2	0.0	0.8	2.3
Total	100.0	100.0	100.0	100.0
Cases	101	113	208	238
Percent in Labor Force	86.6	93.7	96.2	97.7
Percent Unemployed	11.2	1.7	2.6	0.8

Employment of Migrants vs Stayers

The empirical results presented above demonstrate a positive effect of city-ward migration on earnings. It is of interest to examine whether the rise in earnings among migrants was accompanied by greater risk of unemployment as the Harris-Tadaro model suggests. Alternatively, migration may improve employment prospects as Li (1976b) observed for inter-state migration in the United States.

Examination of labor force status reveals that migrants have a lower labor force participation rate than stayers. As shown in Table 7.3 , the proportion of the economically active population among nonmigrants is considerably higher than that of their migrant counterparts. Much of this difference is due to the discrepancy in the proportion of people enrolled in schools. The enrollment rate of recent migrants is four times as high as that of urban stayers. This evidence is consistent with the fact that Taipei is the center of higher education in Taiwan.

A comparison of the economically active population shows that long-term migrants do not have a higher employment rate than stayers. While 1.7 percent of the long-term migrants are unemployed, the corresponding rate is 2.6 percent for urban stayers. It is harder to make comparisons to rural stayers because most persons living on farms tend to report themselves as employed even if there is very little work for them to do. The data suggest that urban-to-Taipei migration has no long term effect on employment. However, a striking difference in the unemployment rate among migrants emerges when duration of residence in the city is taken into account. Recent migrants are characterized by a relatively high unemployment level with about 11 percent of their labor force unemployed. The finding reveals the uncertainty and difficulties in seeking employment among the new arrivals in the city. These data suggest that employment prospects improve significantly with time in the city. This improvement may also be due to (1) changes over time in migrant selectivity (e.g., from selective migration to mass migration), and (2) the biases caused by return and/or onward migration of migrants who were unsuccessful in finding work.

THE COMPARISON OF MIGRANTS AND TAIPEI NATIVES

This section aims to examine economic success of migrants by the standard of city natives. In order to accomplish this task, the labor income of migrants and Taipei natives will be evaluated first by direct comparison (with no or crude control on background variables) and then by regression analyses. We will then investigate differences in the extent and nature of employment between migrants and Taipei natives.

Earnings of Migrants versus Taipei Natives

Utilizing median earnings as a measure, it is found that migrants have higher labor income than Taipei natives. As shown in Table 7.4, those who moved have average monthly earnings of NT$ 4,237 compared to NT$ 3,476 for Taipei natives. This relationship remains for migrants of any length of residence in Taipei. As expected, the difference in median income (between migrants and natives) is especially sharp for long-term migrants (i.e., NT$ 779 in comparison with NT$ 384 for recent arrivals). The data further suggest that previous urban exposure exerts a positive influence upon migrants' earnings. This statement is supported by the fact that urban-to-Taipei migrants earn NT$ 764 more than Taipei natives, while the median labor income of those who moved from rural places exceeds that of nonmigrants in the city by NT$ 471.

Among those with more education, migrants earn slightly less than Taipei natives (NT$ 4,328 versus NT$ 4,455). The data also indicate that recent migrants with more education have a lower median income than nonmigrants in the capital city. Following a longer exposure to Taipei, however, migrants are able to improve their earnings. The median labor income of long-term migrants is NT$ 4,743 as compared to NT$ 4,238 for

TABLE 7.4

MEDIAN MONTHLY EARNINGS OF MIGRANTS AND TAIPEI NATIVES BY EDUCATION

	Education		
Migration Status	All	More Educated	Less Educated
Taipei Natives	3476	4455	2534
All Migrants	4237	4328	3710
Long-term Migrants	4255	4743	3736
Recent Migrants	3860	4238	3362
Urban Origin Migrants	4240	4766	3707
Rural Origin Migrants	3947	4042	3720

Source: 1973 Taiwan Migration Survey

Notes: More educated - Junior high graduates and over.
Less educated - Less than junior high graduates.
Earnings in Taiwan dollars.

recent arrivals and NT$ 4,455 for Taipei natives. These findings suggest that in the long run migrants become well assimilated into the Taipei labor market and their access to high-wage jobs is equal to, or exceeds, that of the city-born. The fact that long-term migrants have higher median earnings than nonmigrants in Taipei also suggests the extra efforts made by migrants themselves (i.e., the effect of migrant adjustment) in addition to the possible influence from return and onward migration.

Table 7-4 also shows that urban background exerts a positive impact upon migrants' earnings. The median income earned by the more educated urban-to-Taipei and rural-to-Taipei migrants is NT$ 4,766 and NT$ 4,042, respectively, as compared to NT$ 4,455 for Taipei natives. This finding seems to imply: (1) among the more educated migrants, those from urban places have received more and/or better quality education than those of rural origin, and (2) previous exposure to an urban way of life facilitates the process of migrant adjustment in the metropolis.

Less educated migrants earn considerably more than Taipei natives (NT$ 3,710 versus NT$ 2,534). This relationship consistently holds even when migrants are grouped into recent and long-term or into urban and rural origin categories. Moreover, while long-term migrants have higher median earnings than recent migrants (NT$ 3,736 versus NT$ 3,362), rural-urban background does not appear to influence migrants' labor income. That is, those who moved from urban places have similar earnings to migrants from rural areas (NT$ 3,707 versus NT$ 3,720). These figures reinforce the earlier interpretation that the less educated migrants probably have made special efforts to successfully adjust to the city environment.

In sum, the direct comparisons reveal that migrants experience no inferiority to lifetime Taipei residents in terms of earnings. In several cases, those who were born and socialized outside Taipei even appear to be in a better position than the city-born. These findings support the research hypothesis that the movers are well assimilated into the urban labor market in Taipei. The possible reasons for some migrants having higher earnings than Taipei natives include such factors as: (1) the economically successful adjustment made by migrants, (2) the return or onward migration of some economic failures, and (3) the more favorable characteristics of migrants compared to natives which were not completely held constant. For instance, the dichotomy of education presented in the direct comparisons is crude and insufficient for assessing the complex relationship between migration and earnings. In addition, other variables such as years of experiences and type of employment may differ between migrants and natives. Therefore, it is advisable to employ a more elaborate technique such as regression analysis for comparing the earnings of migrants and stayers. In this way, detailed and simultaneous controls for earnings determinants other than migration status are possible.

As in the migrant-stayer comparison, the specification of the earnings function is guided by the human capital theory (Becker, 1964; Becker and Chiswick, 1966; Mincer, 1974). The human capital models take length of schooling, level of training, and experience as the basic cause of inter-personal variations in labor income. In this study, the human capital formulation of earnings function is augmented with a set of categorical dummy variables representing migration status.

As argued and evidenced elsewhere (e.g., Mazumdar, 1976; Merrick, 1976; Peek and Antolinez, 1977; Yap, 1976), earnings differentials in an urban economy cannot be explained solely by human capital terms. In addition to the earning capacity of a worker, his location in an institutionally determined part of the labor market is one essential element in accounting for inter-personal variations in earnings. Hence, working status and occupational status are included as dummy variables in the regression analysis.

The earnings functions used to determine the extent of migrant assimilation are of the following form:

$$LnY = a + bS(1) + cS(2) + dS(3) + eS(4) + f(ED) + q(EP) + h^{*}SQR(EP) + i(WK) + j(OC) + D.$$

where:

LnY = natural log of monthly earnings of an individual,

S(i) =	migration status i of the individual [Taipei native (omitted), long-term urban-to-Taipei migrant (i=1), recent urban-to-Taipei migrant (i=2), long-term rural-to-Taipei migrant (i=3), recent rural-to-Taipei migrant (i=4)]
ED =	years of schooling of the individual,
EP =	potential years of work experience of the individual (=age-ED-6),
SQR(EP) =	the potential years of work experience squared,
WK =	working status [nonemployee, i.e., employer or own-account worker (omitted), employee],
OC =	occupational status [blue collar (omitted), white collar],
D =	the disturbance term which includes the effect of natural ability and luck.

The sample used in regression analysis includes 296 cases, 165 of them are migrants and 131 Taipei natives. Each observation is again weighted to adjust for oversampling of married males and recent migrants.

Table 7.5 reports the regression estimates which describe a familiar earnings pattern. The return to education is strongly positive (.053). On average, an increase in schooling by one year would cause an increase in monthly earnings of 5.3 percent. The coefficient of the variable of years of schooling is statistically significant at the one percent level.

The coefficient of the variable of years of experience (0.025) is also significant and with the expected sign. By contrast, the coefficient of the variable of years of experience squared (-0.0004) is insignificant, although it is in the expected direction. The regression coefficients of working status and occupational status are both highly significant and have the expected signs. The estimates show that the white collar worker has an average level of earnings 20 percent higher [i.e., exp (0.18) - 1] than that of the blue collar worker. As far as working status is concerned, the mean labor income of employees is estimated at 85 percent [i.e., exp(-0.16)] of the level of the proprietors (the category of proprietors include employers and own-account workers). This finding is similar to that reported for Malaysia, Peru, Tanzania (Mazumdar, 1976).[11] and the most prosperous (Southern) region of Brazil (Yap, 1976).[12]

----

[11] The findings in these three countries show a wide diversity of earnings among the self-employed and suggest that a substantial proportion of this group performs better than the wage-earners.

[12] Yap found that the self-employed and the employers have an average income level 15 percent higher than that of the employees. However, this pattern was not consistently found in the other two less

TABLE 7.5

EARNINGS FUNCTION IN THE URBAN LABOR MARKET, TAIPEI, 1973
DEPENDENT VARIABLE: Ln (Monthly earnings in N.T.$)

Independent Variables	Regression Coefficients
Migration Status	
Longterm Urban Migrant	0.0737
	(0.900)
Recent Urban Migrant	0.0411
	(0.538)
Longterm Rural Migrant	0.0493
	(0.577)
Recent Rural Migrant	0.1298
	(1.377)
Years of Schooling	0.0528*
	(5.396)
Potential Years of Experience	0.0247*
	(1.707)
Potential Years of Experience Squared	-0.0004
	(0.952)
Working Status: Employees	-0.1589***
	(2.597)
Occupation: White Collar	0.1776***
	(2.877)
Constant	7.5294
R Square	0.203
Sample Size	296

Source: 1973 Taiwan Migration Survey

Notes: The t-ratio is in parentheses.

***Statistically significant at 1 percent level for two-tailed t-test.

*Statistically significant at 10 percent level for two-tailed t-test.

As mentioned earlier, given the substantial increases in earnings associated with migration, it is not likely that migrants are at a competitive disadvantage to their counterparts in the city. Table 7.5 reveals that migrants to Taipei are perfectly assimilated into the urban labor market. Regardless of place of origin and duration of residence in the city, migrants earn as much as lifetime Taipei residents with similar background characteristics. The coefficients of the four migrant groups are all positive but not statistically significant. In other words, when compared with city natives, migrants do not seem to suffer from the fact that their quality of education may have been poorer or from their less frequent exposure to urban life styles. Of course, average migrant earnings might be somewhat lower if one included the economic failures who have returned home or moved onward.

In sum, the regression analysis shows that the earning pattern of migrants is virtually indistinguishable from that of Taipei natives. Given the fact that migrants have derived significant income benefits from moving, the findings imply that the migrant gains have resulted primarily from the higher wages for all jobs in Taipei than in the rest of Taiwan.

Employment of Migrants versus Taipei Natives

A comparison of labor force status (Table 7.6) shows that migrants have a slightly higher labor force participation rate than urban natives (91 percent versus 86 percent). The differences are especially marked when natives are contrasted with long-term migrants (94 percent economically active) and with rural-to-Taipei migrants (96 percent).

Migrants, on the whole, were similar to natives in their absorption into the urban labor market. The figures indicate that those who moved to Taipei have the same unemployment rate as natives (about 5 percent). As expected, duration of residence in the city has a positive influence upon employment. Almost all long-term migrants (98 percent) succeed in obtaining jobs whereas the recent arrivals face some difficulties in finding jobs as indicated by their unemployment rate of about 11 percent. Similar results have also been observed in other countries (Goldscheider, 1983; Oberai, 1977; Peek and Antolinez, 1977: 292-293).

DIFFERENCES IN TYPE OF EMPLOYMENT IN TAIPEI

The above findings suggest that with comparable background characteristics, migrants earn as much as nonmigrants in the capital city. This close similarity in labor income may relate to similarities in employment characteristics between the two groups, but that need not be the case. In order to test this hypothesis, migrants will be compared to lifetime residents in Taipei along various dimensions of employment, such as industry, occupation, and class of worker.

TABLE 7.6

LABOR FORCE STATUS OF MIGRANTS AND TAIPEI NATIVES
(Percentage Distribution)

Labor Force Status	Taipei Natives	Migrants				
			By Duration of Residence		By Place of Origin	
		Total	Long-term	Recent	Urban	Rural
Employed	81.6	86.9	92.1	76.9	84.8	89.7
Unemployed	4.6	4.4	1.6	9.7	3.2	6.1
Student	10.6	7.9	6.3	11.2	12.0	2.4
Others, native labor force	3.2	0.8	0.0	2.2	0.0	1.8
Total	100.0	100.0	100.0	100.0	100.0	100.0
N	183	214	113	101	125	89
Percent in Labor Force	86.2	91.3	93.7	86.6	88.0	95.8
Percent Unemployed	5.3	4.8	1.7	11.2	3.6	6.4

Industry

To examine the industrial composition of the urban labor force, the employed populations are classified into primary, secondary or tertiary industries,[13] with the secondary and tertiary sections further subdivided into the modern and the traditional categories.[14] Since the number of

----

[13] Primary industries are agriculture, forestry, fishing, hunting, and livestock production. Secondary industries are mining and quarrying, manufacturing, construction and electricity, gas, and water supplies. Tertiary industries are commerce, transportation, storage, and communication, service, and others.

TABLE 7.7

PERCENTAGE DISTRIBUTION OF EMPLOYED POPULATION BY INDUSTRY FOR MIGRANTS AND TAIPEI NATIVES

		Migrants				
			By Duration of Residence		By Place of Origin	
Industry*	Taipei Natives	Total	Long-term	Recent	Urban	Rural
Secondary						
Modern	33.4	26.1	24.3	30.4	30.9	20.1
Traditional	14.1	8.4	8.6	7.8	5.3	12.6
Tertiary						
Modern	23.7	34.0	34.9	32.4	33.4	34.7
Traditional	28.8	31.5	32.2	29.4	30.4	32.6
Total	100.0	100.0	100.0	100.0	100.0	100.0
N	173	190	107	83	109	81

*The three-sector grouping of industry into primary, secondary, and tertiary is the same as that used in the Quarterly Report on Labor Force Survey in Taiwan (Taipei, Taiwan: Taiwan Provincial Labor Force Survey and Research Institute).

---

14 This classification is based upon employment status, type of industry, and size of establishment. All government employees belong to the modern sector. Any other respondent whose industrial classification is in the categories of (i) banking, insurance and other financial institutions and (ii) public administration as well as international and foreign institutions is classified as being in the modern sector. Respondents are also considered to be working in the modern sector if they worked in an establishment with seven employees or more. The remainder are classified as being in the traditional sector.

Taipei residents employed in agriculture is trivial, this study ignores the primary sector. The data in Table 7.7 show that migrants differ from Taipei natives in the industrial distribution. The difference is especially pronounced for the tertiary-modern category, where migrants have significantly more participation. In contrast, slightly higher proportions of Taipei natives as opposed to migrants are employed in both the traditional and modern secondary industries. These results reinforce the argument that migrants are more likely to be absorbed by the commercial and service jobs.

The figures in Table 7.7 further suggest that duration of stay in the city does not have a strong impact upon the industrial composition of migrants. Both recent movers and long-term migrants show a close similarity in all categories except for a slight difference in the secondary-modern industries. However, place of origin appears to be a factor affecting the industrial distribution of migrants. Migrants from urban places have a greater share of population in the secondary modern employment than those who moved from rural places. The reverse pattern is found for the proportions in the secondary - traditional industry. There is no substantial difference in the percentage of tertiary-modern and tertiary-traditional categories between urban-to-Taipei migrants and rural-to-Taipei migrants.

When the more educated and less educated persons were examined separately, the same pattern of industrial distribution persists for both Taipei natives and migrants. Among the more educated, lifetime city residents had a higher proportion holding secondary-modern jobs than migrants and a lower proportion in tertiary-modern industries. For the less educated group, migrants are more likely than natives to be employed in the tertiary-traditional sector. This finding is consistent with the view that low education migrants tend to obtain jobs in the informal service sector.

Occupation

There are also differences in the occupational distribution between migrants and urban natives. As shown in Table 7.8, migrants are more likely to be engaged in the white-collar occupations (especially professional, clerical, and sales) than nonmigrants in Taipei. Similarly, migrants appear to be underrepresented in blue-collar and miscellaneous jobs. This finding confirms the widely held notion and previous empirical results (e.g., Oberai, 1977; Yap, 1976) which suggest that in the urban labor market, migrants are not being discriminated against. Compared to lifetime residents in the city, migrants appear to have somewhat more access to high-ranking occupations.

The data further indicate that migrants experience upward occupational attainment after moving to the city. Facing a new environment and numerous uncertainties, recent migrants generally accept low-paid occupations such as service or clerical jobs. With increasing length

TABLE 7.8

PERCENTAGE DISTRIBUTION OF EMPLOYED POPULATION BY OCCUPATION FOR MIGRANTS AND TAIPEI NATIVES

		Migrants				
			By Duration of Residence		By Place of Origin	
Occupation	Taipei Natives	Total	Long-term	Recent	Urban	Rural
Professional	4.3	12.3	12.2	12.6	11.1	13.8
Managerial	6.1	7.5	8.7	4.9	4.8	11.0
Clerical	12.1	17.6	16.5	20.4	23.6	9.7
Sales	20.8	25.9	28.7	19.4	24.5	27.9
Service	7.0	5.7	2.6	12.6	4.8	7.0
Laborer	45.4	30.1	31.3	27.2	29.6	30.6
Miscellaneous	4.3	0.9	0.0	2.9	1.6	0.0
Total	100.0	100.0	100.0	100.0	100.0	100.0
N	175	191	107	84	110	81

of residence in the city, migrants succeed in upgrading their occupational status. Evidence of this is the proportion in blue-collar (i.e., service and laborer) occupations for long-term migrants is smaller than that of recent arrivals (34 percent versus 43 percent). Conversely, there is a significant increase in certain areas of white-collar employment, namely, sales and managerial jobs (37 percent for long-term migrants as compared to 24 percent for recent migrants).

Differences in the occupational composition of urban-to-Taipei migrants and those who move from rural areas also exist. However, the relationship is not in the expected direction. The rural-to-Taipei migrants seem to be superior to urban-to-Taipei migrants in the sense of their having a higher proportion in the high ranking white-collar occupations (i.e., professional and managerial, 25 percent versus 16 percent) but a smaller share in the lower-paid jobs such as clerical (10 percent versus 24 percent). However, there is little discrepancy in the percentage of those in sales or blue-collar occupations between the two migrant groups.

Although migrants were somewhat better educated than Taipei natives, the same pattern in the relationship emerges, when education is controlled. That is, migrants appear to be better off than Taipei natives in

TABLE 7.9

PERCENTAGE DISTRIBUTION OF EMPLOYED PERSONS BY CLASS OF WORKER FOR MIGRANTS AND TAIPEI NATIVES

	Taipei Natives	Migrants				
			By Duration of Residence		By Place of Origin	
		Total	Long-term	Recent	Urban	Rural
Employer	6.8	5.7	6.9	2.9	4.8	6.9
Own Account Worker	22.7	27.9	31.8	18.5	22.9	34.3
Government Employee	11.5	17.0	16.4	18.4	19.5	13.7
Private Employer	52.6	49.1	44.9	59.2	52.8	44.4
Unpaid Family Worker	6.4	0.3	0.0	1.0	0.0	0.7
Total	100.0	100.0	100.0	100.0	100.0	100.0
N	178	192	108	84	110	82

terms of occupational achievement. For the more educated group, the differences are especially sharp in their share of professional jobs (15.6 percent vs 6.5 percent). For the less educated group, migrants exceed natives in their share of the white-collar occupations as a whole.

Class of Worker

Comparisons between migrants and Taipei natives reveal that migrants are not inferior to the city-born in terms of class of worker (Table 7.9). The classification of class of worker in this section consists of five major categories: (1) employers, (2) own-account workers, (3) government employees, (4) private employees and (5) paid family workers. Compared with lifetime residents, migrants in Taipei comprise a slightly higher proportion of the self-employed (i.e., employers and own-account workers) and a greater percentage are employed as government employees. Conversely, migrants appear to be underrepresented in the categories of private employees and unpaid family workers.

The data in Table 7.9 further indicate that those who recently moved to Taipei are more likely to work as employees than long-term migrants. This observation may reflect the fact that recent migrants often lack capital and experience to operate their own enterprises. As a result, the share of enterprise owners (i.e., employer and own-account worker) for recent migrants is less than that of Taipei natives (21 percent versus 30 percent). It appears that, as migrants continue living in the capital city, they are able to accumulate wealth and experience and consequently shift from working as employees to employers and own-account workers. In fact the percentage of long-term migrants who are own-account workers is higher than that of Taipei natives (32 percent versus 23 percent, $t=1.70$) while the proportion of employers reveals no difference between the two groups. However, this difference could also be due to differences in the opportunities available to migrants in different time periods.

It should also be noted that migrants from rural areas are more likely to be self-employed than those from urban areas. However, it is theoretically unclear why migrants with a rural background appear to be better in this respect than their counterparts from urban places. Conversely, the proportion of government and private employees of those who moved from the countryside is less than that of urban-to-Taipei migrants. These findings clearly suggest that migrants from both urban and rural places are able to compete favorably with Taipei natives for white-collar jobs.

In conclusion, the analyses of employment characteristics show no significant difference in absorbing migrants and lifetime residents into the urban labor market in Taipei. Among the employed population, there is no evidence that migrants are more concentrated in the traditional sector industries. In terms of occupation and working status, movers fare as well as, or, in some cases, better than their native counterparts. The findings suggest that migrants are not being discriminated against in obtaining work opportunities and have shared with Taipei natives the favorable labor market conditions.

THE PROCESS OF MIGRANT ADJUSTMENT

As noted earlier, migrants in Taipei appear to be better off due to their movement. Analyses presented in the preceding section further suggest that migrants are not inferior to Taipei natives in earnings or in employment patterns. These findings imply that the improvement enjoyed by migrants may mainly result from the general changes in labor market conditions which favor the receiving area over the sending area.

Nevertheless, this conclusion does not exclude the possible effects of migrant adjustments on migrants gains. In fact, to the extent that migrants are initially in a disadvantageous position relative to Taipei natives (e.g., having a lower quality of education), the process of adjustment has over time eliminated the detrimental effects of migrant background. Otherwise, migrants would have been inferior to similar lifetime Taipei residents in earnings or employment patterns.

Retrospective job histories for the 133 male migrants who were respondents permit us to explore the changes in migrants' jobs over time in Taipei.[15] As found in Surabaya, Indonesia (McCutcheon, 1978:87), however, only a minority (22 percent) of the respondents reported that they had changed jobs at least once in Taipei. The following analyses are based on this small sample.

Examination of the data (not shown) reveals that the first job was a full-time job for most people. Only 13 percent of the sample took their first jobs on a part-time basis. The statistics also show that almost one-half of the respondents kept the first job for less than one year and 29 percent for more than five years. As far as the reason for leaving the first job is concerned, two out of three respondents expressed dissatisfaction with that job in terms of low wages or bad working conditions. These job changes appear to reflect a normal process of economic adjustment of migrants.

Table 7.10 shows the cross-classification of various employment characteristics of migrants at the time of their arrival in Taipei with those at the time of interview. For type of industry, the results show that almost two-thirds of the cases remained unchanged. Among those who changed industrial sectors, there is a tendency for more men to shift from the secondary sector to the tertiary sector (23 percent) than the reverse (14 percent). Due to small sample size, however, the difference is not statistically significant. This slight net inflow from the secondary industries to the tertiary industries was also found, but to a lesser extent, in the previous cross-sectional comparisons between long-term migrants and recent migrants. The greater degree of change in the longitudinal data than the cross-sectional data is partly due to the fact that the analysis based on retrospective data is consistent with the trend of overall development in Taipei, as discussed in chapter 4, which has been towards growth for the tertiary sector.

The cross-tabulation of occupational status reveals that about one-half of the cases maintained the same occupational class for both the first and the current jobs (Table 7.10). Among those who changed occupational status, the proportion of upward mobility cases (31 percent) is almost double that of downward mobility cases (17 percent), although the difference is not significant using the McNemar Test. Consequently, there

----

[15] Retrospective data on first job in Taipei were not gathered for husbands of female respondents.

TABLE 7.10

EMPLOYMENT CHARACTERISTICS FOR MIGRANTS WHO HAVE CHANGED JOBS AT LEAST ONCE IN TAIPEI
(Percentage Based on Totals)

First Job in Taipei	Current Job at the Time of Interview	
Industry (N = 25)		
	Secondary	Tertiary
Secondary*	20.5	22.7
Tertiary	13.6	43.2
(Insignificant based on McNemar Test)		
Occupation (N = 27)		
	Blue Collar	White Collar
Blue Collar	33.3	31.3
White Collar	16.6	18.8
(Insignificant based on McNemar Test)		
Working Status (N = 29)		
	Employees	Self-Employed
Employees	55.7	30.8
Self-employed	5.8	7.7
(Significant at 10% level based on McNemar Test)		

* Including three cases in the primary industry.
**Including employees and own-account workers

Note: Weighted to compensate for the fact that recent migrants were sampled at two times the rate of long-term migrants.

The McNemar Test for the significance of changes (Bishop, et al., 1975:285-287; Siegal, 1956:63-67.)

is a net increase in the proportion in white-collar occupations. As mentioned by Zimmer (1973:445), these changes are influenced both by the aging of the population and by the overall change in occupational composition in the receiving area. In the past three decades of rapid economic growth, there has been a general upgrading of occupational positions over time in Taipei. Nevertheless, the changes observed here by analyzing the retrospective data confirm the earlier findings from the cross-sectional comparison of current occupation between recent and long-term migrants which is free from the two influences mentioned above. Hence, a consensus is that, on average, migrants show successful adjustment by moving away from the blue-collar jobs to more favorable positions in the occupational structure as their duration of residence in Taipei increases. At least, the data suggest that migrants have participated in the general upgrading of occupations in Taipei.

For class of worker, the data indicate that 63 percent of the respondents had the same class for both the first and the current jobs (Table 7.10). Among those who changed class, a majority went from employees to enterprise owners (employers or own-account workers). Less than 6 percent of the entire sample lost the status of proprietors and became employees. In this case, the McNemar Test shows that the difference is statistically significant at the 10 percent level. Again, these changes are affected both by the aging of the population and by the overall change in the structure of working status in Taipei. However, unlike the case of Surbaya, Indonesia (McCutcheon, 1978), the analysis of retrospective data in this study reinforces the earlier conclusion drawn from the contrast of current status between recent and long-term migrants. That is, with increasing length of residence, migrants have improved their working status because they have been able to change from employees with no proprietorship at all to enterprise owners.

It is unfortunate that, due to the extremely small number of cases, the investigation cannot yield further insights. However, the analysis of retrospective data on occupational class and employment status lends support to the influence of migrant adjustment. The general pattern in the economic adjustment of migrants is to shift from the blue-collar occupations to higher ranking positions and to improve working status through the change from employees to business owners.

SUMMARY AND CONCLUSION

Given that migrants in Taipei are better off due to their movement, this study further points out that the movers are in economic positions which are at least as good as Taipei natives in terms of earnings and employment characteristics. In other words, migrants have been well assimilated into the urban labor market of Taipei and are not being discriminated against. These findings imply that a portion of the migrant gains (compared to stayers at origin) is the result of the more favorable labor market conditions in Taipei than in the rest of Taiwan. The evidence from Taiwan is consistent with findings for several other developing countries in Asia and Latin America (Tsay, 1979:19-26). In particular, this research has reached a conclusion similar to that for the investigation for Brazil (Yap, 1976). Moreover, as evidenced by the upgrading of employment characteristics, migrants also seem to have made an economically successful adjustment in the city. The experience of migrants in Taipei reinforces the general consensus of the "migrant adjustment" studies in Bogata, Seoul, Surabaya and Tehran (Goldscheider, 1983). That is, the adjustment of migrants to urban conditions is not a serious problem, at least in the countries under study.

The success of migrants to Taipei is partly a result of the favorable labor market conditions in Taipei. As pointed out in earlier chapters, Taipei population growth has coincided with rapid economic growth. This growth has created certain urban problems which were discussed in chapter 4. However, it is clear from this analysis that migrants will continue to come to Taipei as long as it provides them with an opportunity to improve their economic status. Policies to discourage the city-ward migration, in the interest of alleviating urban problems, which do not take account of the migrant experience in the metropolitan center, are unlikely to be effective. Accordingly, the implication is that development strategies which influence the regional distributions of job creation and opportunities for advancement in the expanding industrial, manufacturing, and service sectors should be adopted in order to prevent further population concentration in Taipei.

Finally, migrants to Taipei are not only highly selective in terms of education as compared to stayers, but are also similar (or slightly superior) to Taipei natives in length of schooling (Tsay, 1979: 83-86, 137-140). The city-ward migration to Taipei appears to be a "pull type" rather than a "push type". Taipei Municipality is the cultural and economic center to which the more ambitious people are drawn for further education and enterprise opportunities. In the city, the similarity in educational attainment between migrants and natives has undoubtedly expedited the process of migrant adjustment and eased the possible discrimination against migrants. Therefore, the ability to generalize from the findings of this study may be limited to countries where educational selectivity of migrants and their similarity to the city-born in schooling exist. To the extent that education influences employment and earnings, it is possible that other settings which differ in this aspect might yield different results on the economic outcomes of individual migrants.

8

Living Conditions and Social Life of Migrants in the City

In the previous chapter, we showed that migrants to Taipei were successful in terms of employment status, occupation, and income in comparison with native born city residents. In this chapter, we will address the issues related to the adjustment of migrants to the living conditions and social life of the city. After brief discussions of previous studies and the methodology of studying migrant adjustment, we will examine in turn, housing and living conditions, social interaction, and measures of individual modernity.

The findings in chapter 7 were consistent with those of several other studies which have shown that migrants to major cities in developing countries adjust well to the economic conditions in these cities. However, there have been fewer studies of the adjustment of migrants to the living conditions and social life of the city. Some of the earlier studies, such as those of Abu Lughod (1961) and Lerner (1958) have pointed to problems in the social adaptation of migrants to city life, while some more recent studies, such as those by McCutcheon (1983) and Clark (1983) show that migrants adjust relatively well, although they may lag behind natives in the acquisition of permanent housing and consumer durables.

Migrants may be slower than urban natives to adapt to urban life for two reasons. First, as newcomers to the city, they lack access to land and other resources which natives may have inherited or been given by relatives in the city. It may be relatively easier for a migrant who possesses the proper education and job skills to find a job in a growing labor market than it is to acquire land and a house in an already crowded city. Secondly, many migrants maintain contact with people who remain behind at their place of origin and may even continue to own land there. This limits the resources they have available for the purchase of a home in the city and, to the extent that they envision eventually returning to the place of origin, it also reduces their willingness to make a major investment in the city. This divided loyalty between the place of origin and the city of destination also limits the involvement of migrants in the political and social life of the city.

Despite these disadvantages, migrants may adjust well because of positive personal qualities which are related to their decision to move. We have shown earlier that migrants to the city tend to be younger and better educated than those who chose to remain at the places of origin. They were also younger and somewhat better educated than native-born Taipei residents. Overall the migration to Taipei tends to conform to the model of selectivity of migration between relatively developed areas of advanced societies. Wilson (1986), in extending Zelinsky's (1971) theory of the mobility transition, argues that while migrants from less developed areas to more developed areas tend to have levels of education which are intermediate between those at origin and destination, those between developed areas tend to be better educated than those in both origin and destination areas. Migration to Taipei fits this model because the majority of the migrants come from other urban areas and because those from rural areas have had access to education.

If migrants are selected on education and work related skills, they may also possess superior qualities such as intelligence, motivation, and competitiveness which are not so easy to measure. These could lead them to make more effort to acquire housing and to participate in social life than the native born.

During the 1970s a series of studies of migrant adjustment were conducted at Brown University (Brown, 1983; Clark, 1983; Corno, 1983; Hendershot, 1978; McCutcheon, 1983; Tirasawat, 1978) using methods similar to those of the Taipei study.[1] In each of these studies, comparisons were made between recent migrants, long-term migrants and natives who were of similar age and sex. Most of these studies addressed common questions of living conditions and social life.

Several of the studies found that the housing conditions of migrants were not as good as those of urban natives. For example, Clark (1983) found that migrants to Seoul were roughly similar to urban natives in terms of overall housing and neighborhood quality, once age, education and income were controlled. However, migrants were less likely to be homeowners and lived in housing units with more persons per room than natives. Somewhat similar results were found for migrants to Tehran (Brown, 1983). However migrants to Bangkok and Manila were disadvantaged relative to natives on most housing measures (Tirasawat (1978) and Hendershot (1978), respectively). In Surabaya, Indonesia, McCutcheon (1983) observed that migrants' housing conditions were not

----

[1] Actually, the Taipei survey served as the first test of a core questionnaire which was incorporated into most of these studies. Those by Hendershot (1978) and Tirasawat (1978) used different questionnaires, but attempted to follow the same methodology in the analysis.

significantly different from those of natives. Only those migrants to Surabaya who came from farm backgrounds or who were self-employed with low incomes had poorer housing conditions; these were offset by migrants with higher levels of education who enjoyed better living conditions than the average urban native.

Fewer previous studies in less developed countries settings have looked at subjective measures of housing such as residential satisfaction. Yi (1985) compared satisfaction with housing and location between migrants and nonmigrants in a 1978 survey in Taichung. He found that nonmigrants were slightly more satisfied than migrants, but the difference was not statistically significant when other variables were controlled.

A useful measure of wealth and level of living which can be easily obtained in surveys in less developed countries is a simple count of the number of a specified list of modern consumer objects which a family owns (Freedman, 1975). These lists vary by country, but tend to include both relatively inexpensive items such as clocks and radios, and expensive items such as refrigerators, and automobiles or motorcycles. In Seoul and in Tehran, recent migrant possessed fewer of these items than did natives, while long-term migrants were intermediate between these two groups (Clark, 1983 and Brown, 1983, respectively). However, in Surabaya, where the average number of expensive items was considerably less than Seoul or Tehran, the recent migrants were not significantly different from the urban natives (McCutcheon, 1983).

In terms of social life, these studies tended to show that migrants had less involvement in urban social institutions and fewer people whom they could call upon in times of need than natives (Goldscheider, 1983). Migrants tended, however, to be similar to urban natives in terms of modern attitudes.

In general the previous studies of adjustment of migrants to living conditions and social life in cities have not been as unanimous in their findings as the studies of economic adjustment. While migrants approach natives on some dimensions in some cities, they fall behind natives on other dimensions and in other cities. Before we see how migrants to Taipei compare to those in other countries, we need to first deal with some of the methodological issue involved in such studies.

METHOD OF STUDYING MIGRANT ADJUSTMENT

The main method of analysis will be to compare two groups of migrants, the recent migrants (who moved within the preceding five years) and the long-term migrants (who have been in the city for at least five years) to otherwise comparable persons who were born in the city. This is essentially the same method which was employed in chapter 7. This approach requires the following assumptions:

(1) that migrants of different durations of residence were similar to

> one another at the time of their arrival in the city; (2) that neither the urban setting to which they are adjusting nor the area of origin from which they have come has changed significantly in terms of the dimensions being investigated; and (3) that the survivors of the initial migration cohorts who remain in the city to the time of the survey are representative of the entire group of migrants who entered the city in that cohort. (Speare, 1983, p. 27)

It is clear that these assumptions can not be strictly met in the case of Taipei or in any city undergoing growth and development. The conditions in Taipei in the early 1950s when some of the first migrants in this study moved to the city were clearly not the same as those in the early 1970s when the most recent migrants came to the city. However, while economic and living conditions have changed, the change in Taipei has been gradual. Taiwan has not experienced the political upheavals which occurred in Korea and Indonesia nor the extremely rapid growth which characterized Tehran prior to the study and contributed to the revolution which occurred soon after the study.

Throughout the analysis in this chapter we shall attempt to control statistically for the most important differences between the three comparison groups which could affect the results. Using Multiple Classification Analysis, age, sex, education, and marital status will be statistically controlled so that the results reported will be equivalent to those of samples which were matched on these characteristics.

While we will continually discuss migrant adjustment on each of the dimensions studied, it is important to keep in mind that the term "adjustment" is a relative one. What appears to be adjustment to one observer may not appear so to another. Although we have taken the urban natives as the standard for comparison, it is possible that the urban natives are not well adjusted to city life on some of the dimensions which we shall be studying. In fact on some dimensions, the migrants may appear to be "better adjusted". In a sense, what we are really doing is defining success on each of several dimensions using our own standards and comparing migrants and natives on those dimensions. For example, we shall assume that the more consumer durables a family has, the more successful it is on that dimension. It is quite possible that by a different standard, the possession of automobiles and motorcycles by urban families, for example, could be viewed as disadvantageous and a higher value placed on the use of public transportation as a measure of urban adjustment.

HOUSING AND LIVING CONDITIONS

Most migrants are young adults who come to the city to improve their employment and income opportunities. It is only reasonable that their initial energies will be directed towards acquiring a job, developing a career, and maximizing income. This is likely to lead to long hours of work and to less emphasis on finding satisfactory living conditions. We should, therefore, expect to find that the housing and living conditions of recent migrants are not on par with those of residents who were born and grew up in the city. It is not clear, however, how long this situation is likely to continue. We should expect to find that long-term migrants have better housing and living conditions than recent migrants, but perhaps not as good as the urban natives. In the following analysis, we shall attempt to determine which aspects of housing and living conditions differ between migrants in the expected direction, which differ little among groups, and which show some other relationship.

One of the most striking differences between migrants and natives is homeownership. Without taking account of differences in age, marital status and other variables, only 27 percent of recent migrants are homeowners compared to 66 percent of the urban natives, while long-term migrants are intermediate with 45 percent homeowners. Controlling for age, sex, marital status, education and whether or not the respondent (or spouse) was the head of household, these percentages change slightly so that 31 percent of the recent migrants are homeowners compared to 63 percent of natives (see Table 8.1).

Urban natives have a clear advantage in homeownership. Many of them are not heads of household, but live in homes owned by their parents, which is much less likely for migrants. Even those migrants who moved with and continue to live with their parents are more likely to rent than natives. The same is true for married couples who have their own household. Among married heads of households, 27 percent of the recent migrants and 38 percent of the long-term migrants own their house compared to 64 percent of the urban natives who are heads, controlling for age and education (See Table 8.2). The fact that long-term migrants are only about half way between the recent migrants and natives on this measure suggests that homeownership is one adjustment to urban life which requires a long time.

There are also differences in some of the measures of housing quality, although these are not as great as those in housing tenure (See Table 8.1). Migrants were less likely to live in homes which had separate kitchens, private toilets or separate rooms for entertaining guests (equivalent to living rooms). On all of these measures, the long-term migrants were intermediate between the recent migrants and the natives. Similar results were obtained when the analysis was limited to married heads of households (See Table 8.2). These results support the view that some migrants are willing to compromise their living conditions while they secure their economic position in the city.

Migrants do not appear to be disadvantaged on other housing measures. Almost all households had running water and the percentages with water were identical for all three groups. The high proportions with running water are in sharp contrast with other Asian cities such as Manila and Jakarta where only a minority of urban residents had running water in their homes in 1973.

TABLE 8.1

COMPARISON OF HOUSING CONDITIONS OF MIGRANTS AND URBAN NATIVES

	Recent Migrants	Long-Term Migrants	Urban Natives	Statistical Significance
Percentage homeowners	31	45	63	.001
Percentage with separate kitchens	79	82	92	.003
Percentage with running water	96	96	96	n.s.
Percentage with private toilet	78	81	88	.04
Percentage with separate room for entertaining guests	76	78	90	.003
Average floor space (sq. meters)	71.0	83.3	78.3	n.s.
Interviewer's Rating of Housing Conditions	3.0	3.2	3.1	n.s.
Residential satisfaction	15.3	15.2	15.3	n.s.
Number of consumer durables	6.2	7.0	7.1	.009
No. of Cases*	146	202	198	

Notes: Based on 1973 Taiwan Migration Survey. Recent migrants are those who moved to Taipei in the five years preceding the survey. Long term migrants moved to Taipei more than 5 years before the survey. Urban natives were born in Taipei and continued to live there. All measures adjusted for age, sex, marital status, education and whether respondent (or spouse) was head of household.

TABLE 8.2

COMPARISON OF HOUSING CONDITIONS OF MARRIED MIGRANTS AND URBAN NATIVES WHO ARE HEADS (OR SPOUSES OF HEADS) OF HOUSEHOLDS

	Recent Migrants	Long Term Migrants	Urban Natives	Statistical Significance
Percentage homeowners	27	38	64	.001
Percentage with separate kitchens	79	80	96	.03
Percentage with running water	94	93	96	73
Percentage with private toilet	76	80	87	n.s.
Percentage with separate room for entertaining guests	72	72	91	.03
Average floor space (sq. meters)	58.1	69.7	71.3	n.s.
Interviewer's rating of housing conditions (1 = very good to 5 = poor)	2.9	3.1	3.0	n.s.
Residential satisfaction	15.2	15.2	14.9	n.s.
Number of consumer durables	6.0	6.7	7.1	.09
No. of Cases*	70	122	49	

N.S. = not significant at p = .1 level.
*Maximum numbers. For some variables these numbers were reduced by non-response.
See notes to Table 8.1.

There was also no difference between migrants and natives in terms of the average amount of floor space, the interviewer's rating of the housing conditions, or the respondents' self evaluation of residential satisfaction. This was the case for both the total sample and for married heads of households (See Tables 8.1 and 8.2). While it is possible that some of the migrants might have been satisfied with poorer quality housing than the natives because these migrants were using a rural basis of comparison, there is no reason to expect that the interviewer's standards for rating housing conditions would have been different for migrants and natives since the same interviewers interviewed all groups.

The final measure of living conditions is the possession of consumer durables. This measure is based on original research by Freedman (1975) and has proved to be useful because it relies on concrete questions with unambiguous answers. In our survey, respondents were asked about their possession of 13 specific items. These included both minor items such as radios, rice cookers and gas stoves which were owned by the majority of Taipei households, and larger items such as air conditioners, automobiles and color televisions which were less frequently owned. Recent migrants owned about six of these items, on the average, compared to long-term migrants and urban natives who had about seven items (see Table 8.1). Although recent migrants had incomes similar to those of Taipei natives, they were were somewhat behind the natives in their accumulation of these items.[2] Presumably this is because it takes time to acquire these items and the recent migrants have had comparable incomes only since their move.

The mixed picture which these results convey is similar to that from the other studies of migrant adjustment in Asia (Goldscheider, 1983). In most other settings, a difference in homeownership was observed. Where it was not observed, such as in the Surabaya study, there was a considerable amount of temporary squatter settlement which was technically classified as homeownership. There is very little such temporary construction in Taipei.[3] In other measures, the Taiwan survey agrees with most of the others in showing that recent migrants are disadvantaged in terms of some living conditions, but that these conditions improve with duration of residence in the city.

----

[2] Again, the results were similar when limited to married heads, although the level of significance was lower due to the smaller number of cases (See Table 8.2).

[3] According to interviewers' ratings, less than three percent of the housing units were of temporary construction.

INTERACTION WITH RELATIVES

Migrants, by leaving their community of origin, increase their distance from those friends and relatives who do not also move. While most migrants who come to the city have some friends and relatives who are already there, and they tend to be joined by others later, many friends and relatives do not move and this leads to some loss in social interaction. In this section we shall examine the extent to which this is true.

Migrants do not differ significantly from natives in having living parents or parents-in-law. Because the norms for Chinese family relationships are strongly patriarchal, we have shown the husband's family separate from the wife's family for married respondents. For 76 percent to 86 percent of the married couples at least one of the husband's parents is living and for 94 percent to 99 percent, the wife's parents are living (see Table 8.3). Presumably the higher proportion of surviving parents for wives is due to the difference of about 3 years in age at marriage between men and women which makes the husband's parents older than the wife's parents, on the average. These data show that most migrants and natives have the opportunity to interact with parents and the opportunity is the same for migrants and natives except for the distance involved.

There are, however, substantial differences between migrants and Taipei natives in whether or not they are living with their parents. Despite considerable modernization, the norm of continuing to live with the husband's parents after marriage is widely adhered to in Taiwan. In fact, it has been argued that because of increased survival and greater incomes, the incidence of extended family living may be greater in Taiwan in the 1960s and 1970s than it was during "more traditional" times in China (Speare, 1974b). Among Taipei natives, the majority of married couples were living with the husband's parents. However, because of the separation caused by migration in many cases, a much lower proportion of migrants were living with parents (see Table 8.3). Long-term migrants were somewhat more likely to be living with their parents than recent migrants. There are two explanations for this difference. First, some of the long-term migrants moved when they were children and these migrants probably accompanied their parents. Secondly, with greater duration of residence in the city, there has been more time for parents who originally remained behind to join the migrants in the city.

As expected, the incidence of extended families including the wife's parents is much lower than that for husband's parents. Nevertheless, the same relationship to migration status holds with the urban born being most likely to live with wife's parents, the long-term migrants intermediate, and the recent migrants least likely (see Table 8.3). Similar relationships also hold for whether single persons live in the same household as their parents.

For those who do not live in the same household with parents, migration also results in significant differences in the frequency of interaction. Interaction was measured on a scale where 1 was most frequent (several times a week) and 5 was least frequent (less than 4 times

TABLE 8.3

COMPARISON OF INTERACTION WITH RELATIVES FOR MARRIED MIGRANTS AND URBAN NATIVES

	Recent Migrants	Long-Term Migrants	Urban Natives	Statistical Significance
Parents of Male R or Husband				
At least one living	83	76	86	n.s.
Percent living in same house*	25	30	68	.001
Interaction scale (1 = most frequent)**	3.9	3.5	2.7	.005
Parents of Female R or Wife				
At least one living	99	96	94	n.s.
Percent living in same house*	4	8	9	n.s.
Interaction scale (1 = most frequent)**	4.1	3.9	2.8	.001
Percentage giving money to any relative	82	70	68	n.s.
Percentage receiving money from any relative	33	26	23	n.s.
Number of brothers or sisters***	8.8	9.0	8.9	n.s.
Percentage of brothers or sisters who live in Taipei	24	39	69	.001
No. of cases (maximum)	95	157	105	

See notes to Table 8.1.

*Among those with at least one living parent.
**Among those with at least one living parent who are not living with parent(s).
***Includes brothers and sisters of spouse.

a year). As shown in Table 8.3, the average score for frequency of interaction for recent migrants who were not living with parents was about 4, which is equivalent to interaction every 2 or 3 months. Long-term migrants had slightly more frequent interaction than recent migrants, but urban natives had much more interaction. The scores of urban natives were less than 3, which means their average frequency of interaction was more than one or two times per month. The frequency of interaction was about the same for husband's parents and wife's parents indicating that the Chinese family norms apply mainly to living arrangements and not to visits.

In addition to parents, we also examined the availability of brothers and sisters, including those of the spouse. There were no significant differences in the total number of brothers and sisters (and brothers-in-law and sisters-in-law) between migrants and urban natives. However, urban natives had a much higher percent of their brothers and sisters living in Taipei than did migrants. Only about 24 percent of the brothers and sisters of recent migrants lived in Taipei compared to 39 percent for recent migrants and 69 percent for natives (see Table 8.3). As was the case with parents, long-term migrants were both more likely to have moved with their parents and siblings when they were still children and also had more time for their siblings to more independently to Taipei after their move.

Finally, we looked at the flow of financial help among close relatives, either parents or siblings. Since the norm in Chinese society is for grown children to help support their parents, it it not surprising that most migrants and natives sent some money to their parents within the past year. There were also flows from parents to children and to other relatives. The relations with migration status were in a different direction from those for extended living arrangements and interaction. While recent migrants had significantly less opportunity for frequent interaction with close relatives, they were more likely to have either received money or sent money to relatives within the last year. However, these differences were not statistically significant with the relatively small size of the groups in this survey.

Overall, the results of this survey show that migration leads to a considerable reduction in extended family living and frequent interaction with parents and, by imputation, with siblings. This does not, however, reduce the flow of financial aid from children to parents or parents to children, or among other close relatives. It would thus appear that migration results in a change in the opportunities for interaction but not in the adherence to traditional norms of behavior.

URBAN ACTIVITIES AND CONTACT WITH MASS MEDIA

Urban areas offer a range of recreational activities which are different from those available in rural areas. To what extent do migrants participate in these activities ? Do they participate soon after arrival, or does it take time for them to become interested in these activities and to acquire the disposable income needed to pay for them ?

In the Taiwan Migration Survey, we asked about three such activities: going to movies, eating in restaurants, and taking pleasure trips. On the first two of these, there were no significant differences among recent migrants, long-term migrants, and Taipei natives (see Table 8.4). However, recent migrants were less likely to have taken pleasure trips within the past year than either long-term migrants or Taipei natives. There are several scenic spots around Taipei including both beaches and mountain settings with waterfalls and cable cars which attract large numbers of Taipei residents on pleasure trips and it is easy to make arrangements to travel with tour groups to these areas. However, such trips are more expensive than attending movies or eating in restaurants and it appears that the recent migrants need time to acquire sufficient disposable income to partake in these activities.

Another dimension of urban life is the greater reliance on mass media communications compared to word of mouth communications (Wirth, 1938). We asked about the frequency of contact with newspapers, magazines, radio, and television. Each of these was scored on a scale ranging from 1 (daily) to 5 (never) and the scores were summed to form an index ranging from 4 to 20. The mean scores on this index were about 10 which correspond to an average contact with each form of mass media of about 3 times per week.[4] There were no significant differences among migrants and natives when age, sex and education were controlled. This suggests that use of the mass media in a densely settled country such as Taiwan where radio, television and newspapers are widely available even in rural areas, is more a function of education than of recency of arrival in the city.

MODERN ATTITUDES

Inkles and Smith (1974) found that there were significant differences in attitudes towards life between factory workers and farmers. However, they did not directly study migrants, but could only make inferences about the effect of factory work by comparing new and old workers and factory workers versus other urban workers. McCutcheon (1983) used a modified form of the scale used by Inkles and Smith to test for differences between migrants and urban natives in Surabaya and found no significant differences.

Instead of repeating the Inkles and Smith scale, we have chosen a few questions which relate to traditional norms in Chinese society which we

---

[4] A score of 10 corresponds to an average score of 2.5 for each item. On a particular item, a score of 2 equals several times a week and 3 equals 1 or 2 times a week.

TABLE 8.4

COMPARISON OF ACTIVITIES OF MIGRANTS AND URBAN NATIVES

	Recent Migrants	Long Term Migrants	Urban Natives	Statistical Significance
Frequency of going to movies (1 = most frequent)	3.7	3.6	3.5	n.s.
Frequency of eating in restaurants (1 = most frequent)	3.2	3.2	3.2	n.s.
Whether took any pleasure trips in the past year	23	37	39	.002
Scale of contact with mass media (low scores are most frequent contact)	10.0	9.8	9.7	n.s.
No. of cases (maximum)	146	201	198	

See notes to Table 8.1.

expect might change with increased urbanization and modernization. In particular, we have looked at attitudes towards traditional family relations, reliance on children in old age, and attitudes towards fertility and family planning.

Filial piety is a traditional Chinese norm (Freedman, 1966; Litwak, 1960). Children are expected to feel the strongest bond to their parents and their bond to their spouse is considered to be secondary. To some, the bond among brothers is felt to also be stronger than that among spouses. To examine the extent to which these norms continued into the 1970s, we asked "To whom should a married man feel closest: to his wife, to his brother, or to his parents?" The results, shown in Table 8.5, indicate that about 41 percent of all respondents and one-half of the male respondents still adhered to the traditional norm that the strongest bond should be to the parents. However, the differences between migrants and natives were not significant.

Taiwan still lacks a comprehensive system of pensions or government based support for old age, although government workers and some employees of large businesses have such plans. This means that most individuals must rely either on their own savings or on their children for support in old age. Traditionally, old people lived with their married children and received all of their support from children. To test the extent to which this is still the norm in Taiwan, we asked two questions, one about their expectations to live with their children in old age and another about their expected reliance on children for financial support in old age. Between 30 and 40 percent of the respondents expected to live with their children or to rely upon them for support, but there were no significant differences between the migrants and natives.

Finally, we would expect city residence to lead to more favorable attitudes towards the practice of family planning and to small family size than rural residence. If this is true, the long-term migrants should have more favorable attitudes than the recent migrants. However, the data in Table 8.5 show that there are no significant differences among recent migrants, long-term migrants and Taipei residents in terms of either ideal family size or approval of family planning. All groups view relatively small families (2 to 3 children) as ideal and have generally favorable views of family planning.[5]

In summary, on attitude measures, the migrants, whether recent or long-term, did not differ significantly from the Taipei residents. It is tempting to attribute the lack of difference to the fact that migrants, on the whole, were relatively well educated and the majority came from other urban places. In addition, the fact that distances between urban and rural areas in Taiwan are not great and communication between these areas is

----

[5] The average score was about 1.9 on a continuum of approval of 1 = "very much", 2 = "much", 3 = "not so much" and 4 = "not at all".

TABLE 8.5

COMPARISON OF TRADITIONAL ATTITUDES OF MIGRANTS AND URBAN NATIVES

	Recent Migrants	Long-Term Migrants	Urban Native	Statistical Significance
Proportion who believe a man should feel closer to parents than wife:				
All respondents	41	38	44	n.s.
Males	53	46	53	n.s.
Proportion expecting to live with children in old age	30	36	32	n.s.
Percentage expecting to rely on children for financial support in old age	37	39	37	n.s.
Ideal family size	2.7	2.7	2.8	n.s.
Approval of family planning (1 = highest approval)	1.9	1.8	1.9	n.s.
No. of Cases				

See Notes to Table 8.1.

good, suggests that modern attitudes should diffuse rapidly throughout the island so that even the migrants from rural areas will not differ greatly from the urban natives at time of arrival in the city. While these factors may make some of these findings unique to Taiwan, similar results were obtained in Surabaya (McCutcheon, 1983) where education levels were much lower and communication between urban and rural areas much poorer.

CONCLUSION

In many ways the results presented in this chapter were consistent with those of chapter 7 which showed that migrants were well adjusted to the economic aspects of the urban environment. In chapter 8, we have shown that migrants are similar to natives in many aspects of their living conditions, social life, and modern attitudes. However, there were some differences.

While migrants were similar to urban natives in terms of both their ratings and the interviewer's ratings of overall housing conditions, recent migrants were much less likely to own their homes and they had fewer consumer durables than urban natives. They were also more likely to lack specific housing features such as a separate kitchen, a private toilet and a separate room for entertaining guests. With time in the city, the migrants were more likely to become homeowners and to acquire those housing features which they initially lacked, although differences still continued between recent and long-term migrants on all but consumer durables. It would appear that Taipei natives had an advantage in having been born and grown up in the city in that they could inherit land and housing from parents or be helped by parents in ways in which those with parents outside the city could not be helped. Furthermore, although migrants earnings equalled those of natives once they were in the city, earnings had been lower before moving so that migrants had had less chance to accumulate assets.

Migrants were also at a clear disadvantage in terms of interaction with close relatives, since most of their close relatives remained at the place of origin. This led to differences in extended living arrangements between migrants and natives and to differences in the frequency of visits. However, it did not affect the flow of money between migrants and their parents and migrants, which equalled that between city natives and their parents.

In terms of participation in urban recreational activities, use of the mass media and attitudes, the migrants were just as modern as the urban natives. When these dimensions are taken together with housing and social life, the overall picture is one of successful adjustment on the part of migrants to Taipei.

9

Implications for Policy and Further Research

In the preceding eight chapters we have traced the urbanization of Taiwan and examined the processes of migration and commuting which have accompanied the growth of cities. Throughout this discussion we have pointed to specific aspects of the rural to urban transformation in Taiwan which have enabled Taiwan to avoid the excessive flow of population to the largest cities which has been a problem in many other less developed countries. While Taiwan had some advantages at the onset of its development which most other less developed countries did not have, it has also been able to guide the course of urbanization through direct and indirect government policies. Many of these policies could be adopted by other countries.

In the first part of this chapter we will review the major findings of this study. We will then discuss the implications of these findings for the future development of Taiwan looking at the problems which may lie ahead as Taiwan continues to develop and become more urbanized. We will then turn our attention to the problems faced by countries which are at an earlier stage in their development and ask what can be learned from the Taiwan case. We will suggest policies which might help these countries avoid some of the problems associated with urbanization. Finally, we will make some suggestions about how future research might best serve the needs of these countries in monitoring the process of urbanization and in evaluating the success of particular policies.

SUMMARY OF FINDINGS

Economic Development and Population Growth

The course of urbanization in Taiwan was shaped by trends in economic development and population growth. Following Gold (1986), we have divided the post-war development of Taiwan into three phases. The "take-off" phase from 1949-60 emphasized land reform, agricultural development, monetary stabilization and infrastructure development. The "labor intensive" phase from 1960-73 was a period of rapid growth in manufacturing with an emphasis on the production of goods for export. From 1973 to 1985 where our analysis ends, Taiwan's development has been characterized as being in an "industrial upgrading" phase where it has had to adapt to increasing competition from lower wage countries in export markets and has also had to respond to external threats from the People's Republic of China and the rapid increase in oil prices in the 1970s.

Taiwan was fortunate in the timing of its stages of economic development relative to the timing of its post-war population growth. The early stages of development in the 1950s preceded the entry of the postwar baby boom cohorts into the labor market. By the time these larger cohorts entered the labor force in the 1960s, Taiwan had developed sufficient infrastructure to provide for the rapid growth of labor-intensive manufacturing which could take advantage of the rapid growth in the labor force. By the time Taiwan entered its third phase of development and could no longer continue the rapid expansion of labor-intensive export oriented industries, the fertility declines of the 1960s began to have an effect on reducing the number of entrants into the labor force and keeping the labor supply in line with the declining demand for labor. Taiwan was also fortunate to have political stability throughout all three stages of development and to have a family system which encouraged the pooling of resources to support family enterprises or to sponsor individual family members in new ventures.

The Growth of Cities

The strong emphasis on exports in Taiwan's industrialization favored the location of many industries near the two major ports of Taipei-Keelung in the North and Kaohsiung in the South. In fact, the growth of Kaohsiung, the second largest city, was given particular encouragement in the early 1960s by the establishment of the Kaohsiung Export Processing Zone which provided nearly 50,000 jobs by the end of 1972 (Wu, 1985). However, population growth was not as concentrated in these two cities as it might have been. The existence of a good transportation network linking most of the population the availability of labor throughout the island, and several government policies to encourage the dispersal of industrialization led to the growth of several smaller cities and the retention in rural areas of some people who might have moved to cities had the situation been different.

An examination of the history of urban growth in Taiwan, in chapter 2, showed that urban growth was due both to the multiplication of cities and the growth of the population of existing cities. From 1920 to 1985, the number of cities with populations over 100,000 increased from 1 to 24. The growth of these cities, when measured within constant boundaries, varied from 2.4 to 5.6 percent with the most rapid growth occurring in the period from the end of World War II to around 1970. While growth during this period was high, it never exceeded six percent and about one-half of this growth was due to the high rate of natural increase. Considering the rapid industrialization during this period, the urban growth rate was quite modest.

While much of the urban growth was concentrated in the two major cities of Taipei and Kaohsiung and in the cities and towns surrounding Taipei City, there was also a considerable amount of growth in cities and urban areas elsewhere on the island. Unlike many Asian countries where the capital city has grown much more rapidly than other cities and has become a "primate city", the distribution of cities by size conforms well to the ideal rank-size distribution.

The growth of smaller urban places is hard to measure from official data on townships because many townships contain both urban and rural parts and the official designations of townships as urban or rural have rarely been changed to keep up with the changing character of these areas. However, by using village level units it is possible to construct urban areas which correspond to the international concept of urban localities as densely settled places with urban characteristics. A study of these areas revealed that there was significant growth of both large and small areas, with the intermediate cities, those between 100,000 and 500,000, playing an increasingly important role over time.

By the late 1970s, the intermediate cities had a higher proportion of their labor force engaged in manufacturing than the large cities. The levels of employment and labor utilization in the intermediate cities corresponded to those of the large cities with the exception that a slightly higher percentage of workers in the intermediate cities were classified as inadequately utilized due to low levels of income.

Determinants of Urbanization

The close linkage between economic development and urbanization was demonstrated by an analysis of the determinants of urbanization in chapter 3. Using the urbanized areas, the time series analysis showed that urban growth up to 1973 was strongly related to industrialization, although increases in educational attainment and other institutional factors also played a role. A cross-sectional analysis at two points in time showed that migration to urbanized areas was much more strongly related to the relative size of the tertiary sector than the manufacturing sector. This relationship became stronger between 1972 and 1980 as manufacturing growth slowed and the continued expansion of the tertiary sector played a more important role in continued urban growth. However, unlike most other Asian nations, the growth of the tertiary sector has followed industrialization and been primarily concentrated in modern rather than traditional jobs. In this regard, Taiwan's experience is much more similar to the earlier experience of the United States and Western Europe than to most of the other nations of Asia.

Growth of the Taipei Metropolitan Area

Throughout this century, much of the urban growth has occurred in Taipei City and the surrounding satellite cities in Taipei County which are considered to be part of the Taipei metropolitan area. About one-half of this growth has been due to natural increase, although much of the natural increase has been due to children born to migrants. It was estimated that migration and the natural increase of the migrant population accounted for 82 percent of the growth in the Taipei area from 1968 to 1973. Over time the growth in the metropolitan area has shifted from the center toward the periphery. Since 1970 there has been very little growth in the older districts of the city and the central business district has experienced a small decline in population. The most rapid growth has occurred in the new areas which were annexed to the city in 1968 and in urbanized areas surrounding the city.

The metropolitan area has succeeded in providing employment for the growing labor force in the area and over time there have been substantial increases in the education and income of the work force.

While the continued decentralization of growth in the Taipei Metropolitan Area has helped prevent some of the problems which would have occurred if more of this growth had been concentrated in the central city, there have still been some problems. Government programs of slum removal were successful in eliminating most of the squatter settlements within the city. While some of the residents of these areas were accommodated in public housing projects within the city, the demand for public housing has considerably exceeded the supply. The metropolitan area has serious problems of air and water pollution and is facing problems in providing an adequate water supply to meet the demands of an ever increasing population.

Alternate Responses to a Changing Economy

The role of rural to urban migration in the urbanization of Taiwan stands out clearly in data on net migration for the major cities and counties of Taiwan. Over the 30 year period from 1950 to 1980, there was continuous migration away from the more rural counties and to the major cities. In later years, there was also considerable migration into Taipei county which surrounds the city.

Although migration played an important role in Taiwan's transformation from an economy based on agriculture to one based on industry, migration was not as great as it could have been. Much of the transformation was accomplished through the location of industries in rural areas and through the commuting of workers who kept their residence in rural areas. The role of commuting and rural industrialization was clearly seen in studies of the employment of members of rural households and statistics on the locations of new manufacturing establishments.

Results of a large survey on labor mobility which was conducted near the end of the labor-intensive period of development showed that most of the movement from agriculture to manufacturing was accomplished by new entrants to the labor force and that there was relatively little mobility of persons who had already established a career in farming. The combined effects of rural industrialization, commuting from rural areas, and the migration of mostly young persons near the age of entry into the labor force meant that the disruptions caused by rapid industrialization were minimal.

Determinants of Individual Migration

In 1973, a special survey was conducted which focused on migrants to Taipei City. A major part of the survey dealt with the factors which were influential in causing some people to decide to move to the city and others to remain in rural areas, urban towns and smaller cities. It was hypothesized that the migration decision would be based on the perceived costs and benefits of the move, the extent of social and economic bonds at the place of origin, the degree of satisfaction with the job and place of residence at the origin, and the awareness of opportunities elsewhere. Some of these concepts such as satisfaction and awareness space did not turn out to be very important factors in distinguishing migrants from stayers, although the lack of differences could be partly due to difficulties in measuring these factors. While there was little difference in general awareness of city opportunities, migrants did differ from stayers in the possession of specific job information before they moved.

The analysis also showed that the perceived costs and benefits were important and that migrants and stayers also differed significantly in their bonds at the place of origin. Stayers were much more likely to be employed and if employed to be working for themselves either on a farm or in a small business. Stayers were also more likely to be home owners. Given that most migrants were young, the majority were living with their parents before moving and this was true even if they had been married at that time. One factor which proved significant was the extent to which they had a say in the expenditure of household income. Migrants were much more likely than stayers to be living in a household where a parent made the decisions and the desire for some autonomy appears to have been one of the motives for moving. Overall, the migration to Taipei City appears to be a rational response to the job opportunities offered in the city in relation to the attachments which one has to the job, home and other people at the place of origin.

The Economic Success of Migrants

A comparison of migrants to Taipei with both nonmigrants at the places of origin and native born Taipei residents showed that migrants were generally successful in terms of employment and income. The analysis, which was described in chapter 7, focused on male migrants because there were too few employed female migrants in the 1973 survey.

Migrants had substantial gains in income in comparison with nonmigrants at the places of origin even when adjustments were made for the higher cost of living in the city. These differences were due in part to the fact that migrants had, on average, higher levels of education than the nonmigrants. However, there were still significant differences when education was controlled. While both recent and long-term migrants fared better than nonmigrants, the long-term migrants did better than the recent migrants. This suggests improvement over time in the city, although these results could also be due to return migration of the less successful. Migrants initially had higher levels of unemployment than nonmigrants, but this difference disappeared with length of residence in the city.

The incomes of migrants also compared favorably to the incomes of Taipei natives. Both recent and long-term migrants earned more than the natives. When these comparisons were made separately by level of education, the biggest differences were found for those in the low education category and these differences could be due to longer working hours or harder work on the part of the migrants. Overall, the level of unemployment for migrants was similar to that of natives, although recent migrants had higher unemployment rates suggesting that some of them required time to find an appropriate job.

Migrants were more likely to be employed in the service sector than natives, especially in professional, clerical and sales jobs or as government employees. Since Taipei is the center of government, it draws government workers from throughout the island. Only 22 percent of the migrants reported having changed jobs after coming to the city and the trend in these shifts was from the manufacturing sector to the service sector.

The overall success of migrants to Taipei may be attributed to the fact that Taiwan is a relatively small area with good transportation and communication so that the places of origin are not very isolated from the capital city. There are no major differences in culture or language and the relatively high levels of education of the migrants meant that they were very similar to the city natives at the time they came to the city.

Living Conditions and Social Life of Migrants

Migrants to Taipei were generally well adjusted to life in the city in terms of housing, social life and the adoption of modern attitudes, as indicated by the analysis in chapter 8. However, there were some significant differences. While migrants housing conditions were judged equal to those of residents by their own subjective standards and the standards of the interviewers, migrants were less likely to own their homes. There were also slightly more migrants than natives who lived in homes lacking facilities such as a separate kitchen, private toilet or a room for entertaining guests. However, the number of housing units with these deficiencies was relatively small. Migrants also possessed fewer modern consumer durables which is probably related to the fact that they earned less at the place of origin and thus had less chance to accumulate the money to pay for these.

By virtue of moving away from their original home, many migrants had less interaction with parents and other relatives than the city natives who were still near their original home. However, migrants did participate in urban forms of recreation, make use of the mass media and share modern attitudes to the same extent as the nonmigrants. Thus, in most respects the migrants can be judged to be well adjusted to life in the city.

URBANIZATION POLICY IN TAIWAN

The previous review of findings has presented a picture of a fairly smooth transition from a rural-agrarian society to an urban-industrial society. While any visitor to Taipei would concede that problems of air pollution, traffic congestion, and lack of green space are apparent, the large cities of Taiwan lack the large numbers of unemployed or underemployed persons who are poorly housed and not fully integrated into the modern life of the city which exist in many other Asian cities. While Taiwan was fortunate in the timing of its industrialization in relation to its most rapid population increase, there were several direct and indirect government policies which had a significant effect on the direction of urbanization in Taiwan.

In the early stages of development, land reform and substantial agricultural development enabled many poor farmers to increase their output and earnings. By turning most tenants into small landowners and assisting them in the use of better seeds, fertilizer and marketing, the government encouraged many who might otherwise have moved to the city to remain in rural areas.

Government investment in transportation, rural electrification and other infrastructure also helped to smooth the rural to urban transition. With the rapid industrialization in the 1960s, many plants were able to locate in rural areas near large cities because adequate transportation and utilities service were available. These plants attracted workers who could commute to work from farm households and reduced the number of workers who had to move to cities. Other early policies which favored decentralization were the implementation of the Local Autonomy System in 1950 which gave counties and large cities authority to plan and administer programs for their own development and the encouragement of small businesses (Tsai, 1987).

While many of these policies were indirect in that their main purpose was not related to population distribution, with later development, population decentralization became a more direct policy goal (Tsai, 1987). One of the policies with direct effects on population concentration was the Industrial Estate Development Program. While this program was in existence from the early 1950s, it did not have a major effect on decentralization until 1972-73 when a significant number of rural industrial estates were developed. A somewhat related policy was the establishment of Export Processing Zones in the mid 1960s which directed a significant amount of new industrialization away from Taipei City towards Kaohsiung and Taichung. This helped maintain a balanced growth among the larger cities.

More recent policies are aimed at further improving transportation to encourage balanced growth in all regions of Taiwan. For example rail lines are being expanded to link the East Coast to Taipei in the North and Kaohsiung in the South and the port in Taichung is being expanded to stimulate more export oriented industries in the central region. The government is also continuing to invest in rural infrastructure and development to continue to encourage people to remain in rural areas (Tsai, 1987).

In the next decade or two, Taiwan will face new problems associated with urbanization. If farming is to remain a viable alternative to the rising incomes available from manufacturing and service sector employment, there will have to be further modifications in the land ownership laws to allow for larger farm sizes. This would make it profitable to raise productivity per farm worker by increased use of agricultural machinery. This need not lead to the release of more farm workers to urban areas if the government continues to encourage the development of nonfarm businesses in rural areas and small towns.

As wages continue to rise and the quality of life in Taiwan improves, there is likely to be increased concern for the ill effects of urban growth such as air and water pollution, traffic congestion, and unsightly urban sprawl. Solving these problems will involve costly government programs and regulations on businesses which will cut into profits and slow the rate of economic expansion as measured by production of goods.

Taiwan is heavily dependent upon world trade. From 1980 to 1985, exports accounted for over one-half of total production. This heavy reliance on world trade will continue to focus much of the urban growth on the capital city because it is both the center for government decision making and the center for finance and international trade in the country. While it is essential that this manufacturing activity take place near the center of political and economic power, much of the activity could be diverted to satellite cities which are physically separate from Taipei City, yet close enough for people and goods to be able to travel quickly to the city. A model for this type of development might be the movement of some of the headquarters of large corporations out of the center of New York City to more suburban locations in Connecticut and New Jersey which are only an hour's drive from the city. Since time and not distance is the main constraint, greater decentralization will be possible as developments in the means of transportation facilitate movement at greater speeds. Decentralization will also be facilitated as communication facilities continue to develop and more and more of the business which now requires face to face contact can take place over telephones or computer lines.

IMPLICATIONS FOR OTHER COUNTRIES

As we move closer to the 21st century and the further development and adoption of technology to minimize the costs of transportation and communication over distance, we need to think of the process of transformation from a rural to an urban way of life as separate from the concentration of population in large cities. Most of the great cities of the Western countries developed before the advent of widespread motor transportation and required dense settlement near the center in order for workers to be able to walk to work and for goods to flow easily between manufacturers, wholesalers and retailers. Many of the Asian cities which developed in the middle of the 20th century experienced similar constraints because there was insufficient wealth to provide for adequate motor transportation.

We are now at a point where incomes have risen sufficiently in many of these countries to permit both widespread motor transportation and the necessary communication systems to enable business to be spread out over greater space. It should be possible both for metropolitan areas in less developed countries to achieve lower densities by expanding their geographical area and for urbanization to be spread more evenly throughout smaller cities and towns. It also seems likely that the boundary between urban and rural areas will become blurred as rural areas take on more and more nonagricultural functions. Instead of a sharp boundary between city and countryside, we are likely to see a continuum from high population densities to low densities which relates more to topography than to economic activity.

As the Taiwan experience has illustrated, the decentralization of urbanization can be facilitated by rural development strategies. It is widely recognized that an extensive network of roads, rural electrification, and good communication are as necessary for rural development as are better irrigation, new seeds and fertilizer. In Taiwan, much of this rural infrastructure was in place by the late 1950s and this set the stage for the decentralized type of development which occurred. Many countries such as the Philippines and Indonesia have made significant investments in rural infrastructure in recent years and these countries have experienced significant increases in agricultural production.

Equally important in a decentralized development strategy is the expansion of the manufacturing of consumer products which farm families can buy in exchange for their increased farm production. Since these industries produce products which will partly be sold in rural areas, there is less need for them to be located near the major ports than is the case for export industries.

The dispersal of industrial growth throughout the country can be further encouraged by specific government policies which are aimed at decentralization. One such policy which has worked well in Taiwan is the industrial zone policy. In countries with relatively high rural population densities such as Taiwan, many rural areas and small towns possess sufficient numbers of available workers to meet the needs of several medium sized industrial plants. By buying large parcels of land in these areas and developing the necessary roads and utility service for these zones, the government can make them attractive sites for the location of new industry. The government can further facilitate the development of these industrial zones by setting up local offices which can provide the necessary permits and take care of most of the other government procedures which are necessary for daily business. Failure to provide these services can force industries to remain in or near the capital city in countries where there is frequent need for government permits to conduct business.

There are several ways in which further urbanization can be decentralized, as the Taiwan case illustrates. Where there is a continued need to be near the capital city or other large cities, it should still be possible to divert much of the new growth into satellites which are physically separated but linked by adequate highways, public transportation and communication lines. This permits the preservation of some green space and the limitation of population densities and the accompanying problems of congestion and pollution. Related to this is the further development of public transportation into rural areas to allow people who chose to do so to remain in rural areas and commute to work in cities. A further step is to encourage small and medium sized industries to locate in rural areas.

Countries can also stimulate development of smaller metropolitan areas in lagging regions by investing more in these regions. One of the most effective ways for the government to stimulate regional development is to decentralize part of the government. With continued improvement in transportation and communication, it should become easier to separate different government departments and institutes so that they need not all be in the same city.

In summary, there should be no need for the continued rapid growth of the giant cities of Asia and other less developed countries. Based primarily on the experience of the 1960s and the 1970s, the United Nations (1985) has projected that there will be 22 urban agglomerations with populations in excess of 10 million persons by the year 2000 and that 17 of these will be in Asia or Latin America. While the meaning of these projections depends in part on whether one envisions cities which approximate their current boundaries or large metropolitan areas with several satellite cities, these projections need not be viewed as inevitable. Countries can influence the future course of urbanization through the use of appropriate incentives and the development of the necessary infrastructure to encourage more of the rural to urban transition to take place in places with lower density.

SUGGESTIONS FOR FURTHER RESEARCH

In many countries it is extremely difficult to monitor the process of urbanization over time because the boundaries of urban areas change frequently and it is impossible to measure growth within constant boundaries. This was not a problem in Taiwan because all of the large cities were divided into districts and all of the counties were divided into townships and these units remained constant over time. In the few cases where there had been annexation, whole townships were added to cities and these then became districts so that their growth could continue to be identified separately. It would facilitate the study of urbanization if more countries would collect population data for subcounty units whose boundaries remained relatively fixed over time. Even where annexation did not involve whole subcounty units, it would be possible for researchers to construct reasonable approximations to city boundaries which remained constant over time. Furthermore, such a system of data collection would make it possible to construct metropolitan areas according to international criteria.

Once a consistent set of small geographical areas were adopted, the study of urbanization would be further facilitated by the collection of other data for these same units such as the number and type of industrial establishments and their employment. For example, in the United States data on employment, payrolls and some other statistics are collected from all businesses on a monthly basis and published for each county.

Since many of the government polices which affect urbanization do not have this as one of their goals, it would be helpful to collect periodic data from businesses on the factors which influence their location decisions and how they perceive changes in locational costs over time. This could be done by regular surveys of businesses throughout the country. Such surveys should help to identify unintended consequences of government regulations on the relative costs of operating in different locations.

There is also a need for more information about how households respond to the changing economic choices available to them during the rural to urban transformation. We showed in Taiwan how the results of a series of farm household surveys were useful in demonstrating the increased reliance on off-farm work and commuting over time. However these were small surveys and it is not known how representative they are. Such surveys need to be large enough to allow for disaggregation and designed to be representative of different types of rural communities and different regions of the country.

Finally, there is a need for more longitudinal studies to show how different groups are affected by the rural to urban transition over time.

While panel studies which follow the same people over a 10 to 20 year period would be ideal, it is often very difficult to keep track of people over that length of time and such studies are particularly likely to lose the migrants - the group which is of greatest interest. There are two alternatives which may be more successful. The first is to collect lengthy life histories in surveys and to use these to trace the past responses of individuals and families to societal changes. The second is to identify and follow cohorts of individuals through a series of censuses or large scale surveys. The cohort method requires that certain identifying information such as place of birth, education and race or ethnicity (where relevant) be collected at each census or survey and that the data be tabulated in detail or available for further tabulation.

This volume has presented a detailed analysis of the trends of urbanization in Taiwan and discussed the determinants and consequences of this urbanization. This analysis has pointed to several ways in which Taiwan has avoided problems associated with excessive urban growth and has also identified some questions which still need to be answered. It is hoped that these results will both help policy makers in less developed countries to plan to avoid some of the problems of urbanization in the future and help researchers to focus their energies in directions which will continue to improve our knowledge about urbanization and development.

Appendix A: The 1973 Taiwan Migration Survey

The Taiwan Migration Survey was conducted in July and August 1973 by the Economics Institute of Academia Sinica in collaboration with Brown University. It was designed especially to test some of the elements of the individual level migration theory discussed in chapter 6 and to collect data on the success of migrants in Taipei relative to native born residents of Taipei.

The Taiwan Migration Survey included four samples. Because the samples were small and because age is an important factor which must be controlled in migration analysis, all samples were restricted to persons aged 20-39. The samples were:

1. Migrants who had moved to Taipei City within the last five years (N=159).

2. Other persons who were living in Taipei City. These could be either migrants of more than 5 years duration in the city or persons born in the city (N=451).

3. Persons living in townships and cities outside Taipei City. Since migrants came to Taipei from all parts of Taiwan, these represent the residents at the places of origin of migrants. This sample was divided between rural and urban areas (N=518).

4. Persons living in townships adjacent to Taipei City which had experienced rapid growth and which could be considered to be integrally related to Taipei even though they were not officially part of it (N=44).

All samples were obtained through a three stage sampling plan. In the first stage, 36 sample towns and city districts were selected from a list stratified by type, location, and size of area. For the second stage, lists of all the lins (small areas of approximately 30 households each) were prepared for each township or city district. Random selections were made from these lists to provide a total of 65 sample lins for Taipei City and 95 for the rest of Taiwan (see Table A-1). Once the sample lins had been selected, enumerators were sent to each of the lins to visit each household and obtain a listing of all household members, their age, sex and date of migration to that city or town if they had not been born there. These lists were used to determine eligibility for the interview survey and within Taipei to distinguish the recent migrants from the other residents. In the final stage, 1 in 2 recent migrants to Taipei, 1 in 4 other residents of Taipei and 1 in 6 residents outside Taipei aged 20-39 were selected to be interviewed.

The interviewing was done by 25 students in the Social Sciences at National Taiwan University. Their work was closely supervised by staff members of the Institute of Economics at Academia Sinica and one graduate student who was hired especially for the survey. The interviewing was preceded by training and practice interviewing and interviews were edited in the field to ensure completeness. Interviewers were asked to make at least three calls at each selected respondent's home to try to obtain an interview. No substitutions were allowed. Because some female respondents objected to being interviewed by male college students, each neighborhood was designated as either "male" or "female" and only respondents of the designated sex were selected in that neighborhood. Female interviewers were used in the "female" neighborhoods.

Interviews were completed with 1172 respondents or 83 percent of those who had been selected to be interviewed. The response rate varied from 75 percent for recent migrants to Taipei to 94 percent for residents of rural areas (see Table A-1). About half of the nonresponse was due to respondents who were not at home despite repeat visits by the interviewers. The rest of the nonresponse was equally divided between refusals and all other reasons such as illness or lack of a common language.

The age and sex distribution of the respondents is shown in Table A-2 by type of sample. The "other Taipei residents" have been divided into long-term migrants and persons who were born in Taipei. The residents outside Taipei are separated into city residents, who include residents of all townships with 50,000 or more population, and rural and small town residents, who include residents of all towns of less than 50,000. This dividing line between urban and rural is a crude one which was originally suggested by Yuan (1964). Speare (1974a), however found that townships over 50,000 usually had net in-migration whereas those under 50,000 usually had net out-migration.

Recent migrants tended to be younger than residents in the areas of origin, as is the case in most studies of migration. However, they were not significantly younger than the persons who were born in Taipei. Combining both sexes, we see from Table A-2 that 69.8 percent of the recent migrants were aged 20-29. Because Taipei has had rapid growth for several

TABLE A.1

NUMBER OF SELECTIONS, COMPLETED INTERVIEWS AND COMPLETION RATE BY SAMPLE TAIWAN MIGRATION SURVEY, 1973

	Recent Migrants in Taipei	Other Taipei Residents	Residents of Other Cities and Urban Areas	Rural Areas	Chungho	Total
Number of Towns or City Precincts Selected	6		11	18	1	36
Number of Lins Selected (small neighborhoods)	65		45	45	5	160
Number of Respondents Selected	212	511	296	263	54	1336
Number of Interviews	159	451	270	248	44	1172
Completion Rate (Percent)	75.0	88.3	91.2	94.3	81.5	87.7

TABLE A.2

AGE AND SEX BY TYPE OF SAMPLE

	(Percentage Distribution)					
	Migrants to Taipei		Persons Born in Taipei	Residents outside Taipei		Chungho Special Sample
Age and Sex	Recent	Long-term		Other Cities	Rural and Small Towns	
Age at Interview						
20–24	41.5	18.6	44.4	27.0	20.6	29.5
25–29	28.3	24.5	23.7	23.3	23.0	25.0
30–34	18.9	30.4	16.2	25.2	27.8	22.7
35–39	11.3	26.5	15.7	24.4	28.6	22.7
Sex						
Male	44.0	43.9	56.1	42.2	46.7	43.2
Female	56.0	56.1	43.9	57.8	53.2	56.8
Total	100.0	100.0	100.0	100.0	100.0	100.0
No. of Cases	159	253	198	270	248	44

decades, its population structure is a young one. Long-term migrants were older, on the average, than recent migrants, the majority being over 30. This is not surprising since much of the rural to urban migration in Taiwan occurs around age 22 when men leave the universally required military service and those moving at age 22 or older would be at least 27 before they had lived long enough in the city to be classified as "long-term migrants".

All of the samples are divided fairly evenly between males and females. Most of the differences in the proportions of males and females between different samples are not statistically significant.[1] However, the samples tend to be biased towards including a higher proportion of females than males due to greater nonresponse among males selected for interviewing and also to the many men aged 20 to 22 who were away in the military.

The age at the time of move is shown in Table A-3. Because of the restriction that all respondents had to be 20 years of age or over at the time of interview and because recent migrants were defined as having lived less than 5 years in the city, the earliest age at move for recent migrants in the survey was 15. The majority of the male and female recent migrants moved before age 25, the largest number having moved between 20 and 24. In contrast a large number of the long-term migrants had come as children.

The distribution of places of origin of migrants to Taipei reveals that migration to Taipei is no longer primarily rural to urban migration. Among recent migrants and long-term migrants who moved as adults, 68.3 and 53.8 percent, respectively, had come from other cities or urban towns. Most frequently they had come from places with populations ranging from 50,000 to 250,000. Nearly one-third of the migrants who had come to Taipei as children came from the Mainland of China, most probably between the time when Taiwan was returned to China in 1946 and the fall of the mainland to the communists in 1949. The movement from the mainland was of quite different character than the movement to Taipei from other places within Taiwan and the migrants from the mainland will therefore be excluded from most of the analysis.

----

[1] For independent random samples of about 200 cases each and proportions between 35 and 65 percent, a difference of about 10 percent between samples is needed for the difference to be statistically significant at the 95 percent confidence level.

TABLE A.3

AGE AT MOVE BY SEX FOR RECENT AND LONG-TERM MIGRANTS

	(Percent Distribution)					
	Male Migrants			Female Migrants		
Age	Recent	Long-Term	Total	Recent	Long-Term	Total
0-14	0.0	40.5	24.6	0.0	31.0	19.1
15-19	15.5	23.4	20.3	8.0	26.1	19.1
20-24	40.8	19.8	28.0	43.2	26.8	33.0
25-29	26.8	11.7	17.6	23.9	7.7	13.9
30-34	9.9	0.9	4.4	13.6	4.2	7.8
35-39	4.2	0.9	2.2	5.7	0.0	2.2
Not known	2.8	2.7	2.8	5.7	4.2	4.8
Total Percent	100.0	100.0	100.0	100.0	100.0	100.0
Cases	70	111	182	89	142	230

Bibliography

Abu Lughod, Janet. 1961. "Migrant Adjustment to City Life: The Egyptian Case." *American Journal of Sociology* 67:22-32.

Anderson, Barbara A. 1977. "Who Chose the Cities? Migrants to Moscow and St. Petersburg Cities in the Late Nineteenth Century." Pp. 277-2 95 in Ronald D. Lee (Ed.), *Population Patterns in the Past*. New York, NY: Academic Press.

Annable, J.E. 1972. "Internal Migration and Urban Unemployment in Low-income Countries: A Problem in Simultaneous Equations." *Oxford Paper* 24:399-411.

Arriaga, Eduardo E. 1968. "Components of City Growth in Selected Latin American Countries." *Milbank Memorial Fund Quarterly:* 46 (Part I): 237-252.

Arrigo, Linda Gail. 1980. "The Industrial Work Force of Young Women in Taiwan." *Bulletin of Concerned Asian Scholars* 12(2):25-34.

Barclay, George W. 1954. *Colonial Development and Population in Taiwan*. Princeton, NJ: Princeton University Press.

Beals, R.E., M.B. Levy and L.N. Moses. 1967. "Rationality and Migration in Ghana." *The Review of Economics and Statistics* 49:480-486.

Berry, B.J.L. 1961. "City Size Distribution and Economic Development." *Economic Development and Cultural Change* 9:573-588.

Berry, R.A. 1975. "Open Unemployment as a Social Problem in Urban Colombia: Myth and Reality." *Economic Development and Cultural Change* 23:276-291.

Barnum, H.N. and R.H. Sabot. 1976. *Migration, Education and Urban Surplus Labour: The Case of Tanzania*. Employment Series No. 13. Paris, France: Development Centre of the O.E.C.D.

Becker, Gary S. 1964. *Human Capital*. New York, NY: Columbia University Press.

Becker, Gary S. and B.R. Chiswick. 1966. "Education and the Distribution of Earnings." *American Economic Review* 56:358-369.

Blau, Peter M. and Otis D. Duncan. 1967. *The American Occupational Structure*. New York, NY: John Wiley and Sons, Inc.

Boulier, Bryan, L. 1977. "Discussion: Migrant in the City." Presented at the Annual Meeting of the Association for Asian Studies, New York City, March 1977.

Brown, Lawrence A., and Eric G. Moore. 1970. "The Intra-Urban Migration Process: A Perspective," *Geografiska Annaler* Series B, 52:1-13.

Brown, Lawrence A. 1972. Questionnaire for Survey of Migrants to Columbus, Ohio (mimeo).

Brown, M. Ray. 1983. "The Adjustment of Migrants to Tehran, Iran," in Calvin Goldscheider, *Urban Migrants in Developing Nations*. Boulder, CO: Westview Press, pp. 189-229.

Browning, H.L. and J.P. Gibbs. 1961. "Some Measures of Demographic and Spatial Relationships among Cities," in J.P. Gibbs (ed) *Urban Research Methods*. Princeton, NJ: D. Von Nostrand Co.

Bureau of Budget, Accounting and Statistics. 1983. *The Statistical Abstract of Taipei City*. Taipei, Taipei City Government.

Caldwell, John C. 1968. "Determinants of Rural-to-Urban Migration in Ghana." *Population Studies* 22(4):361-377.

________________. 1969. *African Rural-Urban Migration*. New York, NY: Columbia University Press.

Chen, Cheng. 1961. *Land Reform in Taiwan*. Taipei: China Publishing Co.

Chinn, Dennis L. 1979. "Rural Poverty and the Structure of Farm Household Income in Developing Countries: Evidence from Taiwan." *Economic Development and Cultural Change* 27(2):283-301.

Cho, Lee-Jay and John C. Baver. 1987. "Population Growth and Urbanization: What Does the Future Hold," in R.J. Fuchs, G.W. Jones and E.M. Pernia (Eds.) *Urbanization and Urban Policies in Pacific Asia*. Boulder, CO: Westview Press, pp. 38-64.

Clark, Sarah C. 1983. "The Adjustment of Migrants to Seoul, Korea," in

Calvin Goldscheider (ed.) *Urban Migrants in Developing Nations*. Boulder, CO: Westview Press, pp. 47-90.

Coale, A.J. and E.M. Hoover. 1958. *Population Growth and Economic Development in Low-Income Countries*. Princeton, NJ: Princeton University Press.

Cohen, M.L. 1976. *House United, House Divided: The Chinese Family in Taiwan*. New York, NY: Columbia University Press.

Corden, W.M. and R. Findlay. 1975. "Urban Unemployment, Intersectoral Capital Mobility and Development Policy." *Economica* 42:59-78.

Corno, Robert. 1983. "The Adjustment of Migrants to Bogota, Colombia," in Calvin Goldscheider (ed.) *Urban Migrants in Developing Nations*. Boulder, CO: Westview Press, pp. 141-184.

DaVanzo, Julie. 1976. "Why Families Move: A Model of the Geographic Mobility of Married Couples," Rand Publication R-1972-DOL.

Davis, Kingsley. 1965. "The Urbanization of the Human Population." *Scientific American* 213:40-53; also pp. 5-20 in G. Breese (ed., 1968), *The City in Newly Developing Countries*. Englewood Cliffs, N.J.: Prentice-Hall, Inc.

________________. 1967. "Population Policy: Will Current Programs Succeed?" *Science* 158:730-739.

________________. 1972. *World Urbanization 1950-1970 Vol. II: Analysis of Trends, Relationships and Development*. Berkeley, CA: Institute of International Studies.

Department of Public Housing. 1975. *Taipei Municipal Housing*. Taipei City Government.

Diamond, Norma. 1979. "Women and Industry in Taiwan." *Modern China* 5:317-337.

Duncan, Otis D. and Beverly Duncan. 1955. "Residential Distribution and Occupational Stratification." *American Journal of Sociology* 60:493-503.

Fields, Gary. 1975. "Rural-Urban Migration and Underemployment and Job Search Activity in LDC's." *Journal of Development Economics* 2:165-187.

Findley, Sally. 1977. *Planning for Internal Migration: A Review of Issues and Policies in Developing Countries*. Bureau of Census. ISP-RD-4. Washington, D.C.: U.S. Government Printing Office.

Frank, C.R. 1968. "Urban Unemployment and Economic Growth in Africa." *Oxford Economic Papers* 20(2):250-274.

Freedman, Deborah S. 1975. "Consumption of Modern Goods and Services and their Relation to Fertility: A Study in Taiwan." *Journal of Development Studies* 12(1):95-117.

Freeman, Maurice. 1966. *Chinese Lineage and Society*. London: Athlone.

Freedman, R. and J.Y. Takeshita. 1969. *Family Planning in Taiwan: An Experiment in Social Change*. Princeton, NJ: Princeton University Press.

Freedman, Ronald, L.C. Coombs, M.C. Chang and T.H. Sun. 1974. "Trends in Fertility, Family Size Preference, and Practice of Family Planning: Taiwan 1965-1973." *Studies in Family Planning* 5(9):270-288.

Galenson, Walter. (ed.) 1979. *Economic Growth and Structural Change in Taiwan*. Ithaca, NY: Cornell University Press.

Gallin, Bernard. 1966. *Hsin Hsing, Taiwan*. Berkeley: University of California Press.

Gallin, Bernard and Rita S. Gallin. 1982. "Socioeconomic Life in Rural Taiwan: Twenty Years of Development and Change." *Modern China* 8(2):205-246.

Gibson, C. 1973. "Urbanization in New Zealand: A Comparative Analysis." *Demography* 10(1):71-84.

Ginsburg, N.S. 1955. "The Great City in Southeast Asia." *American Journal of Sociology* 60:455-62.

Gold, Thomas B. 1986. *State and Society in the Taiwan Miracle*. Armank, NY: M.E. Sharpe Inc.

Goldscheider, Calvin. 1983. *Urban Migrants in Developing Nations - Patterns and Problems of Adjustment*. Boulder, CO: Westview Press.

Goldstein, Sidney. 1971. "Urbanization in Thailand, 1947-1967." *Demography* 8(2):205-223.

________________. 1972. "The Demography of Bangkok: A Case Study of Differentials Between Big City and Rural Populations." Research Report No. 7. Bangkok, Thailand: Institute of Population Studies, Chulalongkorn University.

Goodman, Leo A. 1972. "A General Model for the Analysis of Surveys." *American Journal of Sociology* 77(6):1035-1086.

Green, Sarah C. 1978. "Migrant Adjustment in Seoul, Korea: Employment and Housing." *International Migration Review* 9(1)70-81.

Greenhalgh, Susan. 1984. "Networks and their Nodes: Urban Society on Taiwan." *China Quarterly* (September):51-74.

Greenwood, Michael and Douglas Sweetland. 1972. "The Determinants of Migration Between Standard Metropolitan Statistical Areas." *Demography* 9 (4), November, p. 665.

Greenwood, Michael J. 1975. "Research on Internal Migration in the United States: A Survey." *Journal of Economic Literature.* 13:397-433.

Harris, John and Michael P. Todaro. 1970. "Migration, Unemployment and Development: A Two Sector Analysis." *American Economic Review* 60:96-142.

Hauser, Philip. 1967. "Family Planning and Population Programs." *Demography* 4(1):397-416.

________________. 1974. "The Measurement of Labor Utilization." *Malayan Economic Review* 19:1-17.

Heckman, J. 1979. "Sample Bias as a Specification Error." *Econometrica* 47:153-62.

Heise, D.R. 1969. "Problems in Path Analysis and Causal Inference," in Edgar F. Borgatta (ed.), *Sociological Methodology* San Francisco, CA: Jossey Bass, pp. 38-73.

Hendershot, Gerry E. "Migrant-Native Differences in Housing Quality: Manila, 1973." *International Migration Review* 12(1):104-113.

Hermalin, A.I. 1976. "Spatial Analysis of Family Planning Program Effects in Taiwan." Paper presented at the Seventh Summer Seminar in Population, East-West Population Institute, Honolulu, HI.

Herrick, Bruce H. 1965. *Urban Migration and Economic Development in Chile*. Cambridge, MA: The MIT Press.

Ho, Samuel P.S. 1979. "Decentralized Industrialization and Rural Development: Evidence from Taiwan." *Economic Development and Cultural Change* 28(1):77-96.

Hsieh, Chiao-min. 1964. *Taiwan-Ilha Formosa*. Washington, DC: Butterworth's.

Hsing, M.H. 1971. *Taiwan, Industrialization and Trade Policies*. London: Oxford University Press (OECD).

Inkeles, Alex and David H. Smith. 1974. *Becoming Modern: Individual Change in Six Developing Countries*. Cambridge, MA: Harvard University Press.

Jacoby, Neil. 1966. *U.S. Aid to Taiwan*. New York, NY: Praeger.

Jaffe, A. J. and C.D. Stewart. 1951. *Manpower Resources and Utilization*, New York, NY: John Wiley and Sons.

Johnston, J. 1972. *Econometric Methods*. (2nd Ed.). New York, NY: McGraw-Hill, Inc.

Kmenta, Jan. 1971. *Elements of Econometrics*. New York, NY: MacMillian Inc.

Koo, Anthony Y.C. 1968. *The Role of Land Reform in Economic Development: A Case Study of Taiwan*. New York, NY: Praeger.

Kuo, Shirley W.Y. 1983. *The Taiwan Economy in Transition*. Boulder, CO: Westview Press.

Kuznets, S., R. Easterlin, et al. 1957. *Population Redistribution and Economic Growth - United States, 1870-1950*. 3 Vols. Philadelphia, PA: American Philosophical Society.

Lansing, John B. and Eva Mueller. 1967. *The Geographic Mobility of Labor*. Ann Arbor, MI: Institute for Social Research.

Lansing, John B. and James N. Mogan. 1967. "The Effect of Geographic Mobility on Income." *The Journal of Human Resources* 2(4):449-460.

Lau, Lawrence, J. Ed. 1986. *Models of Development: A Comparative*

Study of Economic Growth in South Korea and Taiwan. San Francisco, CA: Institute for Contemporary Studies.

Lee, T.H. and T.H. Sun. 1972. "Agricultural Development and Population Trends in Taiwan Area, R.O.C." Paper presented at seminar on Effects of Agricultural Innovation in Asia on Population Trends, Manila, Feb. 1972.

Lerner, D. 1958. *The Passing of Traditional Society: Modernizing the Middle East.* Glencoe: IL: The Free Press.

Li, Wen Lang. 1976a. "Internal Migration and Regional Development in Taiwan." Pp. 83-102 in Richmond and Kubat (Eds.). *Internal Migration: The New World and the Third World.* Beverly Hills, CA: Sage Publications.

Li, Wen Lang. 1976b. "A Note on Migration and Employment." *Demography 13(4):565-570.*

Liao, Cheng-hung. 1976. "Rural Labor Mobility in Taiwan - A Summary." Taipei: Department of Agricultural Extension, National Taiwan University.

Lin, T.L. and H.H. Chen. 1969. "Rural Labor Mobility in Taiwan." Taipei: Rural Economics Division, JCRR.

Linsky, Arnold. 1969. "Some Generalizations Concerning Primate Cities," in Gerald Breese (ed.) *The City in Newly Developing Countries.* Englewood Cliffs, NJ. Prentice-Hall, Inc., pp. 285-294.

Little, Ian M.D. 1979. "An Economic Reconnaissance," in Galenson (ed.) *Economic Growth and Structural Change in Taiwan.* Ithaca, NY: Cornel University Press, pp. 448-507.

Litwak, Eugene. 1960. "Geographic Mobility and Extended Family Cohesion." *American Sociological Review* 25:385-394.

Liu, Paul K.C. 1973. *Interactions Between Population Growth and Economic Development in Taiwan.* Taipei: The Institute of Economics, Academia Sinica.

_______________. 1974. "A Study on the Allocation of Manpower and Capital Between Industries in Taiwan" (in Chinese). *Academia Economic Papers (Taipei)* 2(2):151-211.

_______________. 1975. "A Study of Urban Definitions for Taiwan" (in Chinese). Taipei, Taiwan: Economic Planning Council, Executive

Yuan.

________________. 1976a. "The Relationship Between Urbanization and Socio-Economic Development in Taiwan." *Conference on Population and Economic Development in Taiwan*. Taipei, Taiwan: The Institute of Economics, Academia Sinica, pp. 617-646.

________________. 1976b. "Chinese Tradition and Fertility Behavior in Taiwan: An Economic View." *Academia Economic Papers (Taipei)* 4(1):51-78.

________________. 1979. "Economic Aspects of Rapid Urbanization in Taipei." *Academia Economic Papers (Taipei)* 7(1):151-187.

________________. 1983. "Trends in Female Labor Force Participation in Taiwan: The Transition Toward Higher Technology Activities." *Academia Economic Papers (Taipei)* 11(1):293-323.

Liu, Paul K.C. and Alden Speare, Jr. 1973. "Urbanization and Labor Mobility in Taiwan." *Economic Essays* 4:165-177. Taipei, Taiwan: Graduate Institute of Economics, National Taiwan University.

Lo, Fu-chen and Kamil Salih. 1987. "Structural Change and Spatial Transformation: Review of Urbanization Asia, 1960-80," in R.J. Fuchs, G.W. Jones and E.M. Pernia (Eds.) *Urbanization and Urban Policies in Pacific Asia*. Boulder, CO: Westview Press, pp. 15-37.

Lowry, Ira S. 1966. *Migration and Metropolitan Growth*. San Francisco, CA: Chandler Publishing Company.

Mabogunje, A.L. 1972. "Urban Land Policy and Population Growth in Nigeria," in S.H. Ominde and C.N. Ejiogo (eds.) *Population Growth and Economic Development in Africa*. London, England: Heinemann, pp. 235-242.

Mazumdar, Dipak. 1976. "The Urban Informal Sector." *World Development* 4(8):655-679.

McCutcheon, Laurie. 1977. "Migrant Adjustment in Surabaya, Indonesia." Unpublished Ph.D. Dissertation. Providence, R.I.: Brown University.

________________. 1978. "Occupation and Housing Adjustment of Migrants to Surabaya, Indonesia: The Case of a Second City." *International Migration Review* 9(1):82-97.

________________. 1983. "The Adjustment of Migrants to Surabaya, Indonesia," in Calvin Goldscheider (ed.) *Urban Migrants in Developing Nations*. Boulder, CO: Westview Press:95-136.

Mehta, Surinder K. 1964. "Some Demographic and Economic Correlates of Primate Cities: A Case for Revaluation." *Demography* 1:136-47.

Merrick, Thomas W. 1976. "Employment and Earning in the Informal Sector in Brazil: The Case of Belo Horizonte." *The Journal of Developing Areas* 10(3):337-354.

Meyer, David. 1984. "Intermediate Cities in the System of Cities in Developing Countries." *Conference on Urban Growth and Economic Development in the Pacific Region*. Taipei: Institute of Economics, Academia Sinica, pp. 141-159.

Mueller, Eva. 1977. "The Impact of Demographic Factors on Economic Development in Taiwan." *Population and Development Review* 3:(1-2):1-22.

Myers, Ramon H. 1986. "The Economic Development of the Republic of China on Taiwan 1965-81," in L.J. Lau *Models of Development*. San Francisco, CA: Institute for Contemporary Studies, pp. 13-64.

Myrdal, G. 1968. *Asian Drama*, Vol. III. New York, NY: Pantheon.

Mincer, Jacob. 1974. *Schooling, Experience, and Earnings*. New York, NY: National Bureau of Economic Research, Columbia University Press.

Nakoska, R.A. and M. Zimmer. 1980. "Migration and Income: The Question of Self-selection." *Southern Economic Journal* 46 (Jan. 80):840-851.

Nelson, Joan M. 1969. "Migrants, Urban Poverty, and Instability in Developing Nations." Occasional Papers in International Affairs, No. 22. Cambridge, MA: Center for International Affairs, Harvard University.

Oberai, Amarjit S. 1977. "Migration, Unemployment and the Urban Labor Market: A Case Study of the Sudan." *International Labour Review* 115 (2):211-223.

Ominde, S.H. and C.N. Ejiogu. 1972. (Eds.) *Population Growth and Economic Development in Africa*. London, England: Heinemann.

Peck, Peter and Pedro Antolinez. 1977. "Migration and the Urban Labor

Market: The Case of San Salvador." *World Development* 5(4):291-302.

Pernia, Ernesto M. 1977. *Urbanization, Population Growth, and Economic Development in the Philippines."* Westport, CT.: Greenwood Press.

Population Census Office. 1982. *An Extract Report on the 1980 Census of Population and Housing, Taiwan-Fukien Area.* Executive Yuan, Republic of China.

Ranis, Gustav. 1979. "Industrial Development," in W. Galenson (ed.) *Economic Growth and Structural Change in Taiwan.* Ithaca, NY: Cornell University Press, pp. 200-262.

Rondinelli, Dennis A. 1985. *Secondary Cities in Developing Countries: Politics for Diffusing Urbanization.* Beverly Hills, CA: Sage Publications.

Rossi, Peter. 1955 *Why Families Move: A Study in the Social Psychology of Urban Residential Mobility.* New York, NY: The Free Press.

Rostow, W.W. 1960. *The Stages of Economic Growth.* London: Cambridge University Press.

Sahota, Gian S. 1968. "An Economic Analysis of Internal Migration in Brazil," *Journal of Political Economy* 76(2): 218-245.

Schnore, L.F. 1961. "The Statistical Measurement of Urbanization and Economic Development." *Land Economics* 37:229-245.

Schwartz, Aber. 1976. "Migration, Age, and Education," *Journal of Political Economy* 84:4 pp. 701-719.

Scitovsky, Tibur. 1986. "Economic Development in Taiwan and South Korea, 1965-81," in L.J. Lau *Models of Development.* San Francisco, CA: Institute for Contemporary Studies, pp. 135-195.

Scott, Maurice. 1979. "Foreign Trade," in W. Galenson *Economic Growth and Structural Change in Taiwan.* Ithaca, NY: Cornell University Press, pp. 135-195.

Selya, R.M. 1974. *The Industrializatin of Taiwan: Some Geographic Considerations.* Jerusalem: The Jerusalem Academic Press.

Shaw, R. Paul. 1974. "A Note on Cost-Return Calculations and Decisions to Migrate," *Population Studies* Vol. 28, (1): 167-169.

_______________. 1975. *Migration Theory and Fact: A Review and Bibliography of Current Literature.* Bibliography Series No. 5. Philadelphia, PA.: Regional Science Research Institute.

Shryock, Henry S., Jacob S. Siegel and Associates. 1971. *The Methods and Materials of Demography.* Washington D.C.: U.S. Bureau of the Census.

Simon, Herbert A. 1957. *Models of Man.* New York, NY: John Wiley and Sons, Inc.

Sjaastad, Larry, A. 1962. "The Costs and Returns of Human Migration," *Journal of Political Economy.* Supplement on "Investment in Human Beings" 70 (Part 2): 80-93.

Smith, Carol A. 1985. "Theories and Measures of Urban Primacy: A Critique" in M. Timberlake (ed.) *Urbanization in the World Economy.* New York, NY: Academic Press.

Somers, Gerald G. (Ed.) 1967. "The Returns to Geographic Mobility: A Symposium." *The Journal of Human Resources* 2(4) (Fall).

Speare, Alden, Jr. 1969. "The Determinants of Rural to Urban Migration in Taiwan." Ph.D. Dissertation University of Michigan, Ann Arbor, MI.

_______________. 1971a. "A Cost-Benefit Model of Rural to Urban Migration in Taiwan," *Population Studies* 25;117-30.

_______________. 1971b. "An Assessment of the Quality of Taiwan Migration Registration Data," *Taiwan Working Paper No. 12.* Ann Arbor, MI: Population Studies Center, University of Michigan.

_______________. 1973. "The Determinants of Migration to a Major City in a Developing Country: Taichung, Taiwan." *Population Papers: Essays on the Population of Taiwan.* Pp. 167-188 Taipei: Acedemia Sinica

_______________. 1974a. "Urbanization and Migration in Taiwan," *Economic Development and Cultural Change* 22(2):302-317.

_______________. 1974b. "Migration and Family Change in Central Taiwan," *The Chinese City Between Two Worlds* by Mark Elvin and G. William Skinner. Stanford, CA: Stanford University Press. pp. 303-330.

_______________. 1977. Book Review of Todaro (1976), *International*

Migration Review 11(2):248.

________________. 1983. "Methodological Issues in the Study of Migrant Adjustment," Calvin Goldscheider (ed.). *Urban Migrants in Developing Nations*. Boulder, CO: Westview Press, pp. 21-40.

________________. P.K.C. Liu, K.S. Hwang, C.L. Tsay and M. Speare. 1975. "A Measurement of the Accuracy of Data in the Taiwan Household Register." *Academia Economic Papers* 3(2):35-74.

________________., Sidney Goldstein and William H. Frey. 1975. *Residential Mobility, Migration and Metropolitan Change*. Cambridge, MA: Ballinger Publishing Co.

________________ and Sidney Goldstein. 1978. "Summary: Comparative Studies of Migrant Adjustment in Asian Cities," *International Migration Review* 9(1):114-116.

Sun, T.H. 1977. "Demographic Evaluation of Taiwan's Family Planning Program." Paper presented at the Eighth Summer Seminar in Population, Population Institute, East-West Center, Honolulu, HI, June-July.

Suzuki, Keisuke. 1980. *Kukan Jinkogaku* (Space Demography), Taimeido.

Suzuki, Keisuke and Toshio Kuroda. 1981. "On the Structure of the Spatial Distribution of Recent Urban Population in Japan." Research Paper Series No. 4, Population Research Institute, Nihou University, Tokyo, October.

Taiwan Provincial Labor Force Survey. 1984. *Quarterly Report on the Labor Force Survey in Taiwan*.

Tandon, B.B. 1978. "Earning Differentials Among Native Born and Foreign Born Residents of Toronto," *International Migration Review* 12(3):406-410.

Temple, Gordon Paul. 1975. "Migration to Jakarta," *Bulletin of Indonesian Economic Studies* 11:1, pp. 76-81.

Thorbecke, Erik. 1979. "Agricultural Development," in Galenson *Economic Growth and Structural Change in Taiwan*. Ithaca, NY: Cornell University Press, pp. 132-205.

Thornton, Arland, Ming-Cheng Chang and Te-Hsiung Sun. 1984. "Social and Economic Change, Intergenerational Relationships, and Family Formation in Taiwan," *Demography* 21(4):475-499.

Thornton, Arland, Hui-sheng Lin, and Mei-Lin Lee. 1987. "Social Change, The Family, and Well Being." Presented at the Conference on Economic and Social Welfare, Academia Sinica, Taiwan.

Tirawawat, Penporn. 1977. "Urbanization and Migrant Adjustment in Thailand." Unpublished Ph.D. Dissertation. Providence, RI: Brown University.

Tirawawat, Penporn. 1978. "Economic and Housing Adjustment of Migrants in Greater Bangkok." *International Migration Review* 9(1):93-103.

Tsai, Hong-Chin. 1982. "Rural Industrialization in Taiwan," Working Paper, Korea Development Institute, Seoul.

Tsai, Hsung-Hsiung. 1987. "Population Decentralization Policies: The Experience of Taiwan," in R.J. Fuchs, G.W. Jones and E.M. Pernia (Eds.) *Urbanization and Urban Policies in Pacific Asia*. Boulder, CO: Westview Press, pp. 214-229.

Tsay, Ching-lung. 1982. "Migration and Population Growth in Taipei Municipality." *Industry of Free China* 57(3):9-25.

Todaro, Michael P. 1969. "A Model of Labor Migration and Urban Unemployment in Less Developed Countries," *American Economic Review* 59:138-48.

_________________. 1971. "Income Expectations, Rural-Urban Migration and Urban Employment in Africa." *International Labour Review* 104(5):387-413.

_________________. 1976. *Internal Migration in Developing Countries: A Review of Theory, Evidence and Research Priorities*. Geneva, ILO.

Tsui, Y.C. and T.L. Lin. 1964. "A Study on Rural Labor Mobility in Relation to Industrialization and Urbanization in Taiwan." *Economic Digest Series* No. 16, JCRR, Taipei.

United Nations, Department of Economic and Social Affairs. 1968. "Methods of Analyzing Data on Economic Activities of the Population." Population Studies, No. 43, New York, NY: United Nations.

United Nations. 1973. *The Determinants and Consequences of Population Trends*, Vol. 1. New York, NY: United Nations.

Urban Planning Department. 1976. *Taipei Regional Plan* (in Chinese). Taipei, Taiwan: Economic Planning Council.

United Nations. 1985. *Estimates and Projections of Urban, Rural and City Populations, 1950-2025: The 1982 Assessment.* Department of International Economic and Social Affairs, New York.

Vapnarsky, Cesar A. 1969. "On Rank-Size Distribution of Cities: An Ecological Approach." *Economic Development and Cultural Change* 17:584-595.

Weller, Robert H., John J. Macisco, and George R. Martine. 1971. "The Relative Importance of the Components of Urban Growth in Latin America." *Demography* 8(2):225-232.

Wilson, Franklin. 1986. "Aspects of Migration in an Advanced Industrial Society." Madison, WI: Center for Demography and Ecology, University of Wisconsin Working Paper 86-25.

Wirth, L. 1938. "Urbanism as a Way of Life." *American Journal of Sociology* 44:3-24.

World Bank. 1987. *World Development Report 1987.* New York, NY: Oxford University Press.

Wolpert, Julian. 1965. "Behavioral Aspects of the Decision to Migrate." *Papers of the Regional Science Association* 15;159-69.

Wu, Rong-i. 1976. "Urbanization and Industrialization in Taiwan: A Study on the Specific Pattern of Labor Utilization." *Proceedings of Conference on Population and Economic Development in Taiwan.* Taipei, Dec. 29, 1975-Jan. 3, 1976.

________________. 1985. "Taiwan's Success in Industrialization." *Industry of Free China* (Nov.):7-22.

Yang, Martin M.C. 1970. *Socio-economic Results of Land Reform in Taiwan.* Honolulu, HI: University Press of Hawaii.

Yap, Lorene. 1972. "International Migration and Economic Development in Brazil." Unpublished Ph.D. Dissertation. Cambridge, MA: Harvard University.

________________. 1975. "Internal Migration in Less Developed Countries: A Survey of the Literature." World Bank Staff Working Paper No. 215. Washington, D.C.: World Bank.

________________. 1976. "Rural-Urban Migration and Urban Underemployment in Brazil." *Journal of Development Economics* 3(3):227-243.

Yi, Chin-Chun. 1985. "Urban Housing Satisfaction in a Transitional Society: A Case Study in Taichung, Taiwan." *Urban Studies* 22:1-12.

Yuan, D. Y. 1964. "The Rural-Urban Continuum: A Case Study in Taiwan," *Rural Sociology* 26:247-60.

Zelinsky, Wilbur. 1971. "The Hypothesis of the Mobility Transition." *Geographical Review* 61:219-249.

Zimmer, Basil G. 1973. "Migration and Changes in Occupational Compositions." *International Migration Review 7(4):437-447.*

Zipf, G.K. 1941. *National Unity and Disunity: The Nation as a Biosocial Organism.* Bloomington, IN: The Principia Press.

For Product Safety Concerns and Information please contact our EU representative GPSR@taylorandfrancis.com Taylor & Francis Verlag GmbH, Kaufingerstraße 24, 80331 München, Germany

Batch number: 10726013

Printed by Printforce, the Netherlands